THEORY AND ANALYSIS of the DYNAMIC STABILITY of MISSILES

by

Robert L. Swaim

Rocket Engineering Series

Wexford Press
2008

TABLE OF CONTENTS

	PAGE
LIST OF SYMBOLS	vi - xiv
INTRODUCTION	1
ANALYSIS	1
THRUST FORCES AND MOMENTS	4
SLOSHING PROPELLANT FORCES AND MOMENTS	5
ENGINE INERTIA FORCES AND MOMENTS	5
AERODYNAMIC FORCES AND MOMENTS	6
REFERENCES	16
APPENDIX A STRUCTURAL DYNAMICS	17
APPENDIX B ENGINE INERTIA	25
APPENDIX C FUEL SLOSHING	35
APPENDIX D AERODYNAMIC FORCES	41
APPENDIX E DETERMINANT ELEMENTS	51

LIST OF ILLUSTRATIONS

FIGURE		PAGE
1	XZ-Plane for Steady Flight	13
2	XZ-Plane for Disturbed Flight	13
3	XY-Plane for Steady Flight	14
4	XY-Plane for Disturbed Flight	14
5	Typical Closed-Loop System	15
6	Elastic Missile in Disturbed Flight in XZ-Plane	18
7	Elastic Missile in Disturbed Flight in XY-Plane	22
8	Control Engine in Disturbed Flight in XZ-Plane	28
9	Control Engine in Disturbed Flight in XY-Plane	30
10	Sloshing Spring-Mass Analogy	37

LIST OF SYMBOLS

SYMBOL		UNITS
$A_{1,1}$, $A_{1,2}$	Coefficients of stability determinant	
$A^{(i)}(t)$	Defined by Eq. (13A)	
a_{ℓ}	Radius of ℓth tank	ft
$B_{1,1}$, $B_{2,2}$	Coefficients of wind disturbances in XY-plane	
b_1, b_2, b_3	Aerodynamic coefficient reference lengths	ft
$C_{1,1}$, $C_{2,2}$	Coefficients of wind disturbance in XZ-plane	
C_D	Total drag coefficient	
$C_D(\ell)$	Local drag coefficient	1/ft
C_L	Total lift coefficient	
$C_L(\ell)$	Local lift coefficient	1/ft
C_{ℓ}	Total rolling moment coefficient	
$C_{\ell}(\ell)$	Local rolling moment coefficient	1/ft
C_m	Total pitching moment coefficient	
$C_m(\ell)$	Local pitching moment coefficient	1/ft
C_n	Total yawing moment coefficient	
$C_n(\ell)$	Local yawing moment coefficient	1/ft
C_y	Total side force coefficient	
$C_y(\ell)$	Local side force coefficient	1/ft
$\bar{C}_f$	Equivalent gimbal friction coefficient	ft-lb-sec
D	Total drag force parallel to x-axis	lb
$D_{1,1}$, $D_{2,2}$	Coefficients of pitch command signal	
$E(\ell)I(\ell)$	Bending stiffness	$lb\text{-}ft^2$
$E_{1,1}$, $E_{2,2}$	Coefficients of yaw command signal	

LIST OF SYMBOLS (CONT'D)

SYMBOL		UNITS
$F(t)$	Arbitrary function of time	
$F_{1,1}$, $F_{2,2}$	Coefficients of roll command signal	
F_x, F_y, F_z	External forces in X, Y, and Z-directions	lb
F_{x_a}, F_{y_a}, F_{z_a}	Aerodynamic forces in X, Y, and Z-directions	lb
$F_{x_{ei}}$, $F_{y_{ei}}$, $F_{z_{ei}}$	Engine inertia forces in X, Y, and Z-directions	lb
$F_{x_{SP}}$, $F_{y_{SP}}$, $F_{z_{SP}}$	Propellant sloshing forces in X, Y, and Z-directions	lb
F_{x_T}, F_{y_T}, F_{z_T}	Thrust forces in X, Y, and Z-directions	lb
$G(\ell)J(\ell)$	Torsional stiffness	lb-ft^2
g	Acceleration of gravity	ft/sec^2
h_ℓ	Depth of fluid in ℓ^{th} tank measured from tank bottom	ft
$h_{o\ell}$	Distance of ℓth rigid mass from tank bottom	ft
$h_{1\ell}$	Distance of ℓth spring-mass from tank bottom	ft
$I_{xx} = \int_0^L \bar{I}(\ell)\, d\ell$	Reduced mass-moment of inertia of missile about x-axis (calculated with all propellant mass removed from missile)	slug-ft^2
$\bar{I}(\ell)$	Reduced mass-moment of inertia of missile about x-axis per unit length along missile (calculated with all propellant mass removed from missile)	$\frac{\text{slug-ft}^2}{\text{ft}}$
I_{yy}, I_{zz}	Reduced mass-moment of inertia of missile about Y- and Z-axes (calculated with control engine mass and mass of sloshing portion of propellant removed and accounting for the C.G. of the rigid mass obtained from the spring-mass analogy)	slug-ft^2
I_{xz}	Product of inertia	slug-ft^2
$I_{o\ell}$	Mass-moment of inertia of rigid portion of tank propellants about an axis through the C.G. of the rigid portion	slug-ft^2

LIST OF SYMBOLS (CONT'D)

SYMBOL		UNITS
$J_1(\overline{\xi}_\ell)$	Bessel Function of first kind	
$K_{A_{xz}}, K_{A_{xy}}, K_{A_{zy}}$	Gains for controller amplifiers	
$K_{I_{xz}}, K_{I_{xy}}, K_{I_{zy}}$	Gains for controller integrators	
$K_{R_{xz}}, K_{R_{xy}}, K_{R_{zy}}$	Gains for pitch, yaw, and roll rate gyros	sec
K_ℓ	Spring constant for ℓ^{th} spring-mass	lb/ft
K'_ℓ	Defined by Eq. (11C)	
$\overline{K}_1, \cdots, \overline{K}_{10}$	Coefficients of Eqs. (37B) and (38B)	
L	External rolling moment (positive clockwise viewed from rear)	lb-ft
L	Length of missile from nose to control engine C.G.	ft
L_a, L_{ei}, L_{SP}, L_T	External rolling moment due to aerodynamic, engine inertia, sloshing propellant, and thrust forces	lb-ft
ℓ	Coordinate of missile length (positive from nose aft)	ft
ℓ_{CG}	Distance from nose to missile C.G.	ft
ℓ_e	Distance from missile C.G. to engine gimbal	ft
ℓ_R	Distance from gimbal to control engine C.G.	ft
M	Total mass of missile minus mass of control engine and minus mass of sloshing portion of propellants (reduced mass) $M = M_t - M_R - \sum_\ell M_{1_\ell} = \int_0^L m(\ell)\, d\ell$	slugs
M	External pitching moment (positive nose up)	lb-ft
M_a, M_{ei}, M_{SP}, M_T	External pitching moment due to aerodynamic, engine inertia, sloshing propellant, and thrust forces, respectively	lb-ft
M_{c_ℓ}	Defined by Eq. (12C)	slug-ft
M_{P_ℓ}	Total propellant mass in ℓ^{th} tank	slugs

LIST OF SYMBOLS (CONT'D)

SYMBOL		UNITS
M_R	Mass of control engine	slugs
M_t	Total mass of missile	slugs
$M_{0\ell}$	Mass of rigid portion of propellant in ℓ^{th} tank	slugs
$M_{1\ell}$	Spring-mass in ℓ^{th} tank	slugs
$\mathcal{M}^{(i)}$	Generalized mass, or mass moment of inertia, of ith mode	slugs or slug-ft^2
$m(\ell)$	Reduced mass distribution along length of missile	$\frac{\text{slugs}}{\text{ft}}$
N	External yawing moment (positive clockwise viewed from above	lb-ft
N_a, N_{ei}, N_{SP}, N_T	External yawing moment due to aerodynamic, engine inertia, sloshing propellant, and thrust forces	lb-ft
P	Total rolling velocity	rad/sec
$p \cong \dot{\phi}$	Disturbance rolling velocity	rad/sec
Q	Total pitching velocity about reduced missile C.G.	rad/sec
$\bar{Q}^{(i)}(t)$	Generalized force, or moment, for i^{th} mode	lbs or lb-ft
$q \cong \dot{\theta}$	Disturbance pitching velocity	rad/sec
$\bar{q}^{(i)}(t)$	Generalized displacement of ith mode	ft or rad
R	Total yawing velocity	rad/sec
$r \cong \dot{\psi}$	Disturbance yawing velocity	rad/sec
$S_1, \cdots, S_6$	Aerodynamic coefficient reference areas	ft^2
s	Laplace Transform complex variable	
T_c	Thrust of control engine	lbs
T_s	Thrust of sustainer engines	lbs

LIST OF SYMBOLS (CONT'D)

SYMBOL		UNITS
$\bar{T}(\ell,t)$	External moment distribution acting about missile elastic axis	$\frac{\text{lb-ft}}{\text{ft}}$
t	Real time	sec
U	Total linear velocity in X-direction	ft/sec
$\bar{U}$	Relative velocity	ft/sec
u	Disturbance velocity in X-direction	ft/sec
$\bar{u}(eG)$	Elastic displacement at engine gimbal point	ft
$\bar{u}_{xz}(\ell,t)$	Local elastic displacement in XZ-plane perpendicular to undeformed elastic axis	ft
$\bar{u}_{xy}(\ell,t)$	Local elastic displacement in Y-direction perpendicular to undeformed elastic axis	ft
$\bar{u}_{zy}(\ell,t)$	Local elastic displacement (rotation) in a plane perpendicular to elastic axis	rad
V	Total linear velocity in Y-direction	ft/sec
v	Disturbance velocity in Y-direction	ft/sec
v_g	Component of wind disturbance velocity (gust & shear) perpendicular to X-axis in XY-plane	ft/sec
W	Total linear velocity in Z-direction	ft/sec
w	Disturbance velocity in Z-direction	ft/sec
w_g	Component of wind disturbance velocity (gust & shear) perpendicular to X-axis in XZ-plane	ft/sec
$w(\ell,t)$	External normal force distribution on the missile	lbs/ft
X	Aerodynamic force in X-direction	lbs
x	Stability axis coordinate	ft
$\ddot{x}$	Missile acceleration in X-direction	ft/sec^2
x_ℓ	Distance to ℓ^{th} spring-mass location from missile C.G. (positive forward)	ft

LIST OF SYMBOLS (CONT'D)

SYMBOL		UNITS
x_0	Gimbal location with respect to inertial axes	ft
Y	Aerodynamic force in Y-direction	lbs
y	Stability axis coordinate	ft
$Y_{\ell_{xz}}$	Displacement of ℓth spring-mass in XZ-plane perpendicular to missile elastic axis	ft
$Y_{\ell_{xy}}$	Displacement of ℓth spring-mass in Y-direction	ft
Y_0	Gimbal location with respect to inertial axes	ft
Y'_{CG}	Missile C.G. location with respect to inertial axes	ft
Z	Aerodynamic force in Z-direction	lbs
z	Stability axis coordinate	ft
z'_{CG}	Missile C.G. location with respect to inertial axes	ft
z_0	Gimbal location with respect to inertial axes	ft
$\alpha \cong \frac{w}{U_0}$	Disturbance angle of attack	rad
α_0	Steady flight angle of attack	rad
$\beta \cong \frac{v}{U_0}$	Disturbance side slip angle	rad
β_0	Steady flight side slip angle	rad
γ	Disturbance flight-path angle	rad
γ_0	Steady flight, flight-path angle	rad
$\delta_{c\theta}$	Input command to control engine actuator for deflection in XZ-plane	rad
$\delta_{c\psi}$	Input command to control engine actuator for deflection in XY-plane	rad
$\delta_{c\phi}$	Input command to the auxiliary roll control	rad
δ_θ	Control engine deflection in XZ-plane	rad
δ_ψ	Control engine deflection in XY-plane	rad

LIST OF SYMBOLS (CONT'D)

SYMBOL		UNITS
θ	Disturbance pitch angle	rad
θ_c	Programmed pitch command signal	rad
$\theta_F = \theta_{RG} + \theta_{PG}$	Total pitch feedback signal	rad
θ_{PG}	Pitch position gyro feedback signal	rad
θ_{RG}	Pitch rate gyro feedback signal	rad
$\xi^{(i)}$	Damping ratio of ith mode	
ξ_ℓ	Damping ratio of sloshing propellant in ℓ^{th} tank	
$\bar{\xi}_\ell$	First root of Bessel equation, $J_1'(\bar{\xi}_\ell) = 0$	
$\xi_{RG_{xz}}, \xi_{RG_{xy}}, \xi_{RG_{zy}}$	Damping ratios for pitch, yaw, and roll rate gyros	
ρ	Air density	$\frac{\text{slugs}}{\text{ft}^3}$
$\sigma^{(i)}(\ell) = -\frac{d\varphi^{(i)}(\ell)}{d\ell}$	Negative slope of the ith mode function	rad/ft
$\sigma^{(i)}(eG)$	Negative slope of ith mode function at gimbal point	rad/ft
$\sigma_{xz}^{(i)}(PG), \sigma_{xy}^{(i)}(PG), \sigma_{zy}^{(i)}(PG)$	Negative slopes of ith mode function at pitch, yaw, and roll position gyro locations	rad/ft
$\sigma_{xz}^{(i)}(RG), \sigma_{xy}^{(i)}(RG), \sigma_{zy}^{(i)}(RG)$	Negative slopes of ith mode function at pitch, yaw, and roll rate gyro locations	rad/ft
$\sigma^{(i)}(X_{\ell B})$	Negative slope of ith mode function at location of tank bottom along elastic axis	rad/ft
$\tau_{P_{xz}}, \tau_{P_{xy}}, \tau_{P_{zy}}$	Time constants for pitch, yaw, and roll position gyro lag filters	sec
$\tau_{f_{xz}}, \tau_{f_{xy}}, \tau_{f_{zy}}$	Time constants for pitch, yaw, and roll controller filters	sec

LIST OF SYMBOLS (CONT'D)

SYMBOL		UNITS
ϕ	Disturbance roll angle	rad
ϕ_o	Steady flight roll angle	rad
ϕ_c	Programmed roll command signal	rad
ϕ_F	Total roll feedback signal	rad
ϕ_{PG}	Roll position gyro feedback signal	rad
ϕ_{RG}	Roll rate gyro feedback signal	rad
$\phi^{(i)}(\ell)$	Mode function for the ith mode	ft or rad
$\varphi^{(i)}(\ell)$	Normalized mode function for i^{th} mode	
$\varphi^{(i)}(eG)$	Normalized mode function evaluated at gimbal	
$\varphi^{(i)}(\ell_R)$	Normalized mode function evaluated at control engine C.G.	
$\varphi^{(i)}(x_\ell)$	Normalized mode function evaluated at ℓ^{th} spring-mass location	
ψ	Disturbance yaw angle	rad
ψ_c	Programmed yaw command signal	rad
ψ_F	Total yaw feedback signal	rad
ψ_{PG}	Yaw position gyro feedback signal	rad
ψ_{RG}	Yaw rate gyro feedback signal	rad
$\omega^{(i)}$	Undamped natural frequency of ith mode	rad/sec
ω_ℓ	Undamped natural frequency of ℓth spring-mass	rad/sec
$\omega_{RG_{xz}}, \omega_{RG_{xy}}, \omega_{RG_{zy}}$	Undamped natural frequencies of pitch, yaw, and roll rate gyros	rad/sec

SUBSCRIPTS:

o	Steady flight condition	
eG	Engine gimbal location	

LIST OF SYMBOLS (CONT'D)

SYMBOL

xz , xy , zy	Coordinate planes

SUPERSCRIPTS:

(i)	i^{th} elastic mode
(j)	j^{th} elastic mode
ℓ	ℓ^{th} liquid propellant tank
e	Term to be corrected for flexibility of aeroelastic modes not included in equations

INTRODUCTION

The general theory and analysis presented in this report represents what is generally referred to as an "exact" analysis of the dynamic stability of an elastic missile. It is exact in the sense that it deals with the complete equations of motion for a general configuration and a general mission, or trajectory. On the other hand, it is approximate inasmuch as the terms used in the equations and the equations themselves are approximate abstractions of real characteristics.

Much of the existing literature dealing with the dynamic stability of large missiles is limited to the analysis of motion in one plane. This is adequate for some purposes; however, boosters having large, winged payloads and very flexible structures could have significant lateral, longitudinal, and torsional coupling through the external aerodynamic forces and moments and through the effects of propellant sloshing and gimballed engine inertia. An analysis including these cross-coupling effects, therefore, is needed. The analysis developed and presented in this report, we believe, is a substantial step in this direction. The method presented is still based on the usual linearizing small perturbation assumptions. The formidable problem of conducting a completely nonlinear dynamic stability analysis is reserved for the analyst who is forced into dealing with nonlinear equations of motion in order to arrive at more accurate analytical results.

ANALYSIS

The approach that is taken herein to derive the general linear equations of motion for a large, flexible missile is outlined below.

The equations are developed for six degrees of rigid-body freedom plus an arbitrary number of elastic degrees of freedom about all three coordinate axes. The missile is assumed to be propelled and controlled by thrust from fixed and gimballed rocket engines; it may carry winged payloads and external aerodynamic surfaces, but they are assumed sufficiently rigid that their own elastic modes need not be coupled to those of the missile body. The missile may contain either liquid or solid propellants and is assumed to be symmetric about the XZ-plane through the longitudinal axis.

The equations of motion are developed for the system in the presence of artificial mathematical constraints; that is, the sloshing portion of the liquid-propellant mass and the mass of the gimballed control engine are subtracted from the total vehicle mass in writing the rigid-body and elastic equations of motion. Correspondingly, the liquid-propellant mode equation is presented for sloshing in a flexible tank, and equations of motion are presented for the control engine about its gimbal point. These artificially uncoupled modes are then coupled together analytically in writing the equations of motion. They, therefore, have elastic and inertial coupling terms between them in the equations of motion. If this artificial uncoupling approach were not taken, the elastic modes would have to be computed with the sloshing propellants and the oscillating control engines included, and this would be a much more difficult approach.

This report released July 1962 by author for publication as an ASD Technical Documentary Report.

The equations are linearized by assuming small perturbations about the steady-state flight condition and are referenced to a right-hand, stability (Eulerian) axes system. The X-axis is aligned initially with the component of relative velocity lying in the XZ-plane (see Figures 1 and 3). The six rigid-body equations can be expressed in the following familiar form (see Ref. 2). (Note that the total mass M_t has been retained in the gravity terms):

$$M\left[\dot{U}+QW-RV\right]=\sum F_x+M_t g\left[-\sin\gamma_0\cos\theta\cos\psi + \cos\gamma_0\sin\phi_0\cos\theta\sin\psi-\cos\gamma_0\cos\phi_0\sin\theta\right] \quad (1)$$

$$M\left[\dot{V}+RU-PW\right]=\sum F_y+M_t g\left[-\sin\gamma_0(\cos\psi\sin\theta\sin\phi-\sin\psi\cos\phi) + \cos\gamma_0\sin\phi_0(\cos\psi\cos\theta+\sin\psi\sin\theta\sin\phi)+\cos\gamma_0\cos\phi_0\cos\theta\sin\phi\right] \quad (2)$$

$$M\left[\dot{W}+PV-QU\right]=\sum F_z+M_t g\left[-\sin\gamma_0(\cos\psi\sin\theta\cos\phi+\sin\psi\sin\phi) + \cos\gamma_0\sin\phi_0(\sin\psi\sin\theta\cos\phi-\cos\psi\sin\phi)+\cos\gamma_0\cos\phi_0\cos\theta\cos\phi\right] \quad (3)$$

$$\dot{P}I_{xx}-(\dot{R}+PQ)I_{xz}+RQ(I_{zz}-I_{yy})=\sum L \quad (4)$$

$$\dot{Q}I_{yy}+(P^2-R^2)I_{xz}+PR(I_{xx}-I_{zz})=\sum M \quad (5)$$

$$\dot{R}I_{zz}+(QR-\dot{P})I_{xz}+PQ(I_{yy}-I_{xx})=\sum N \quad (6)$$

The above equations are based on the yaw-pitch-roll convention with the variables being positive in the directions indicated in Figures 1, 2, 3, and 4; roll is positive clockwise as viewed from the rear. The terms on the right side of the equations are the externally applied forces, moments, and gravity force components along each axis in the perturbed condition. The equations are written about the C.G. based on the reduced missile mass.

The term "rigid-body" as used herein implies no dynamic oscillation about the elastic axis (no elastic modes); however, it may include terms which have been corrected for elasticity to account for that part of the flexibility not completely described by the elastic modes included in the analysis. Any term or symbol used with a zero subscript is evaluated at the "rigid-body," steady-state, flight condition; and when used with a superscript e, the term is to be corrected for static aeroelastic effects (residual flexibility of higher modes not included).

The U, V, W, P, Q, and R components of velocity can be expressed as the sum of a steady-flight value and a disturbance component as follows:

$$U = U_0 + u \qquad P = P_0 + p$$

$$V = V_0 + v \qquad Q = Q_0 + q$$

$$W = W_0 + w \qquad R = R_0 + r$$

At this point, the assumption is introduced that all disturbance angles and velocities are small as well as the steady-state velocity, V_0 , angle of attack, α_0 , and control engine deflection angles. Also, since stability axes are being used, $W_0 = 0$. Substituting the above components into the equations of motion and neglecting small products yields:

$$M\left[\dot{U}_0+\dot{u}+Q_0 w-R_0V_0-R_0 v-V_0 r\right]$$
$$+M_t g\left[\sin\gamma_0+\theta\cos\gamma_0\cos\phi_0-\psi\sin\phi_0\cos\gamma_0\right]=\Sigma F_x \qquad (7)$$

$$M\left[\dot{V}_0+\dot{v}+R_0 U_0+U_0 r+R_0 u-P_0 w\right]$$
$$+M_t g\left[-\psi\sin\gamma_0-\cos\gamma_0\sin\phi_0-\phi\cos\gamma_0\cos\phi_0\right]=\Sigma F_y \qquad (8)$$

$$M\left[\dot{W}_0+\dot{w}+P_0 V_0+P_0 v+V_0 p-Q_0U_0-Q_0 u-U_0 q\right]$$
$$+M_t g\left[\theta\sin\gamma_0+\phi\cos\gamma_0\sin\phi_0-\cos\gamma_0\cos\phi_0\right]=\Sigma F_z \qquad (9)$$

$$(\dot{P}_0+\dot{p})I_{xx}-(\dot{R}_0+\dot{r}+P_0Q_0+P_0 q+Q_0 p)I_{xz}$$
$$+(Q_0R_0+Q_0 r+R_0 q)(I_{zz}-I_{yy})=\Sigma L \qquad (10)$$

$$(\dot{Q}_0+\dot{q})I_{yy}+(P_0^2+2P_0 p-R_0^2-2R_0 r)I_{xz}$$
$$+(P_0R_0+P_0 r+R_0 p)(I_{xx}-I_{zz})=\Sigma M \qquad (11)$$

$$(\dot{R}_0+\dot{r})I_{zz}+(Q_0R_0+Q_0 r+R_0 q-\dot{P}_0-\dot{p})I_{xz}$$
$$+(P_0Q_0+P_0 q+Q_0 p)(I_{yy}-I_{xx})=\Sigma N \qquad (12)$$

The external forces and moments in the above equations are due to the thrust, aerodynamic, sloshing-propellant, and control-engine-acceleration-reaction (inertia) forces in the respective X, Y, and Z directions.

$$\Sigma F_x = F_{x_T}+F_{x_{SP}}+F_{x_{ei}}+F_{x_a} \qquad (13)$$

$$\Sigma F_y = F_{y_T}+F_{y_{SP}}+F_{y_{ei}}+F_{y_a} \qquad (14)$$

$$\Sigma F_z = F_{z_T}+F_{z_{SP}}+F_{z_{ei}}+F_{z_a} \qquad (15)$$

$$\Sigma L = L_T+L_{SP}+L_{ei}+L_a \qquad (16)$$

$$\Sigma M = M_T+M_{SP}+M_{ei}+M_a \qquad (17)$$

$$\Sigma N = N_T+N_{SP}+N_{ei}+N_a \qquad (18)$$

THRUST FORCES AND MOMENTS

From Figures 2 and 4 it can be determined that

$$F_{x_T} = T_s \cos\left[a_o + \left(\frac{\partial \bar{u}_{xz}}{\partial \ell}\right)_{eG}\right] \cos\left(\frac{\partial \bar{u}_{xy}}{\partial \ell}\right)_{eG}$$

$$+ T_c \cos\left[a_o + (\delta_\theta)_o + \delta_\theta + \left(\frac{\partial \bar{u}_{xz}}{\partial \ell}\right)_{eG}\right] \cos\left[(\delta_\psi)_0 + \delta_\psi + \left(\frac{\partial \bar{u}_{xy}}{\partial \ell}\right)_{eG}\right] \quad (19)$$

$$F_{y_T} = -T_c \sin\left[(\delta_\psi)_0 + \delta_\psi + \left(\frac{\partial \bar{u}_{xy}}{\partial \ell}\right)_{eG}\right] \cos\left[(\delta_\theta)_0 + \delta_\theta + \left(\frac{\partial \bar{u}_{xz}}{\partial \ell}\right)_{eG}\right]$$

$$- T_s \sin\left(\frac{\partial \bar{u}_{xy}}{\partial \ell}\right)_{eG} \cos\left(\frac{\partial \bar{u}_{xz}}{\partial \ell}\right)_{eG} \quad (20)$$

$$F_{z_T} = -T_s \sin\left[a_o + \left(\frac{\partial \bar{u}_{xz}}{\partial \ell}\right)_{eG}\right] \cos\left(\frac{\partial \bar{u}_{xy}}{\partial \ell}\right)_{eG}$$

$$- T_c \sin\left[a_o + (\delta_\theta)_o + \delta_\theta + \left(\frac{\partial \bar{u}_{xz}}{\partial \ell}\right)_{eG}\right] \cos\left[(\delta_\psi)_0 + \delta_\psi + \left(\frac{\partial \bar{u}_{xy}}{\partial \ell}\right)_{eG}\right], \quad (21)$$

where $\left(\frac{\partial \bar{u}_{xz}}{\partial \ell}\right)_{eG}$ and $\left(\frac{\partial \bar{u}_{xy}}{\partial \ell}\right)_{eG}$ are the slopes of the elastic axis due to elastic deflections in the XZ and XY-planes, evaluated at the gimbal point. These equations can be simplified to:

$$F_{x_T} \cong T_s + T_c \quad (22)$$

$$F_{y_T} \cong -T_s \left(\frac{\partial \bar{u}_{xy}}{\partial \ell}\right)_{eG} - T_c \left[(\delta_\psi)_0 + \delta_\psi + \left(\frac{\partial \bar{u}_{xy}}{\partial \ell}\right)_{eG}\right] \quad (23)$$

$$F_{z_T} \cong -T_s \left[a_o + \left(\frac{\partial \bar{u}_{xz}}{\partial \ell}\right)_{eG}\right] - T_c \left\{a_o + \left[(\delta_\theta)_o + \delta_\theta + \left(\frac{\partial \bar{u}_{xz}}{\partial \ell}\right)_{eG}\right]\right\} \quad (24)$$

Also,

$$L_T \cong 0 \quad (25)$$

assuming only one control engine with the gimbal point on the elastic axis and intermittent reaction jet or vernier roll control.

$$M_T \cong -\ell_e \left\{T_s \left(\frac{\partial \bar{u}_{xz}}{\partial \ell}\right)_{eG} + T_c \left[(\delta_\theta)_0 + \delta_\theta + \left(\frac{\partial \bar{u}_{xz}}{\partial \ell}\right)_{eG}\right]\right\} + (T_s + T_c)\, \bar{u}_{xz}(eG) \quad (26)$$

$$N_T \cong \ell_e \left\{T_s \left(\frac{\partial \bar{u}_{xy}}{\partial \ell}\right)_{eG} + T_c \left[(\delta_\psi)_0 + \delta_\psi + \left(\frac{\partial \bar{u}_{xy}}{\partial \ell}\right)_{eG}\right]\right\} - (T_s + T_c)\, \bar{u}_{xy}(eG) \quad (27)$$

SLOSHING PROPELLANT FORCES AND MOMENTS

As mentioned earlier, artificially uncoupling the modes results in external forces and moments due to sloshing propellants. These are covered more fully in Appendix C and presented here for convenient reference.

$$F_{x_{SP}} \cong \sum_{\ell} - M_{I_{\ell}} \, (\dot{U}_0 + \dot{u}) \tag{28}$$

$$F_{y_{SP}} \cong \sum_{\ell} - K_{\ell} \, Y_{\ell_{xy}} \tag{29}$$

$$F_{z_{SP}} \cong \sum_{\ell} - K_{\ell} \, Y_{\ell_{xz}} \tag{30}$$

$$L_{SP} \cong 0 \quad \text{(assuming no circular sloshing)} \tag{31}$$

$$M_{SP} \cong \sum_{\ell} K_{\ell} \, Y_{\ell_{xz}} \, X_{\ell} + \sum_{\ell} M_{c_{\ell}} \ddot{Y}_{\ell_{xz}} + \sum_{\ell} \frac{M_{I_{\ell}} Y_{\ell_{xz}}}{M_t} \, (T_s + T_c - D) \tag{32}$$

$$N_{SP} \cong \sum_{\ell} -K_{\ell} \, Y_{\ell_{xy}} \, X_{\ell} + \sum_{\ell} -M_{c_{\ell}} \ddot{Y}_{\ell_{xy}} + \sum_{\ell} - \frac{M_{I_{\ell}} Y_{\ell_{xy}}}{M_t} \, (T_s + T_c - D) \tag{33}$$

ENGINE INERTIA FORCES AND MOMENTS

The control-engine inertia forces are derived in Appendix B and are presented here for convenience:

$$F_{x_{ei}} \cong - M_R \, (\dot{U}_0 + \dot{u}) \tag{34}$$

$$F_{y_{ei}} \cong - M_R \Big\{ \dot{V}_0 + \dot{v} - (L - \ell_{CG}) \ddot{\psi} + \ell_R \ddot{\delta}_{\psi} + \sum_i \Big[\varphi_{xy}^{(i)}(eG) - \ell_R \, \sigma_{xy}^{(i)}(eG) \Big] \ddot{\bar{q}}_{xy}^{(i)} \Big\} \tag{35}$$

$$F_{z_{ei}} \cong - M_R \Big\{ \dot{w} + (L - \ell_{CG}) \ddot{\theta} + \ell_R \ddot{\delta}_{\theta} + \sum_i \Big[\varphi_{xz}^{(i)}(eG) - \ell_R \, \sigma_{xz}^{(i)}(eG) \Big] \ddot{\bar{q}}_{xz}^{(i)} \Big\} \tag{36}$$

$$L_{ei} \cong 0 \tag{37}$$

$$M_{ei} \cong - M_R \, (L - \ell_{CG}) \Big\{ \dot{U}_0 \alpha_0 + \dot{w} + (L - \ell_{CG}) \ddot{\theta} + \ell_R \ddot{\delta}_{\theta} + \sum_i \Big[\varphi_{xz}^{(i)}(eG) - \ell_R \, \sigma_{xz}^{(i)}(eG) \Big] \ddot{\bar{q}}_{xz}^{(i)} \Big\} \tag{38}$$

$$N_{ei} \cong -M_R (L - \ell_{CG}) \Big\{ \dot{V}_0 + \dot{v} - (L - \ell_{CG}) \ddot{\psi} + \ell_R \ddot{\delta}_\psi + \sum_i \left[\varphi_{xy}^{(i)}(eG) - \ell_R \sigma_{xy}^{(i)}(eG) \right] \ddot{\bar{q}}_{xy}^{(i)} \Big\} \tag{39}$$

AERODYNAMIC FORCES AND MOMENTS

For small disturbances from the steady-flight condition, the aerodynamic forces and moments can be expressed approximately as a sum of first order terms in a Taylor series expansion about the steady-flight values:

$$\begin{aligned} F_{x_a} \cong X_0 &+ \left(\frac{\partial X}{\partial u}\right)_0 u + \left(\frac{\partial X}{\partial v}\right)_0 v + \left(\frac{\partial X}{\partial w}\right)_0 w + \left(\frac{\partial X}{\partial p}\right)_0 p \\ &+ \left(\frac{\partial X}{\partial q}\right)_0 q + \left(\frac{\partial X}{\partial r}\right)_0 r + \left(\frac{\partial X}{\partial \dot{u}}\right)_0 \dot{u} + \left(\frac{\partial X}{\partial \dot{v}}\right)_0 \dot{v} + \left(\frac{\partial X}{\partial \dot{w}}\right)_0 \dot{w} \\ &+ \left(\frac{\partial X}{\partial \dot{p}}\right)_0 \dot{p} + \left(\frac{\partial X}{\partial \dot{q}}\right)_0 \dot{q} + \left(\frac{\partial X}{\partial \dot{r}}\right)_0 \dot{r} + \sum_i \left(\frac{\partial X}{\partial \bar{q}_{xz}^{(i)}}\right)_0 \bar{q}_{xz}^{(i)} \\ &+ \sum_i \left(\frac{\partial X}{\partial \bar{q}_{xy}^{(i)}}\right)_0 \bar{q}_{xy}^{(i)} + \sum_i \left(\frac{\partial X}{\partial \bar{q}_{zy}^{(i)}}\right)_0 \bar{q}_{zy}^{(i)} + \sum_i \left(\frac{\partial X}{\partial \dot{\bar{q}}_{xz}^{(i)}}\right)_0 \dot{\bar{q}}_{xz}^{(i)} \\ &+ \sum_i \left(\frac{\partial X}{\partial \dot{\bar{q}}_{xy}^{(i)}}\right)_0 \dot{\bar{q}}_{xy}^{(i)} + \sum_i \left(\frac{\partial X}{\partial \dot{\bar{q}}_{zy}^{(i)}}\right)_0 \dot{\bar{q}}_{zy}^{(i)} \end{aligned} \tag{40}$$

$$F_{y_a} \cong Y_0 + \Delta Y \tag{41}$$

$$F_{z_a} \cong Z_0 + \Delta Z \tag{42}$$

$$L_a \cong L_0 + \Delta L \tag{43}$$

$$M_a \cong M_0 + \Delta M \tag{44}$$

$$N_a \cong N_0 + \Delta N \tag{45}$$

where the Δ terms are expansions similar to the one in Eq. (40). The derivative terms are evaluated in Appendix D in terms of aerodynamic coefficients and stability derivatives. The rigid-body equations are now complete.

It now remains to derive the equations of motion for the elastic modes in all three planes, control engine motion about the gimbal point, and propellant sloshing in two planes (it is assumed that there is no circular sloshing about the missile axis). These equations are developed in Appendices A, B, and C, respectively, and are written below for easy reference.

$$\ddot{\bar{q}}_{xz}^{(i)} + 2\xi_{xz}^{(i)} \omega_{xz}^{(i)} \dot{\bar{q}}_{xz}^{(i)} + (\omega_{xz}^{(i)})^2 \bar{q}_{xz}^{(i)} = \frac{\bar{Q}_{xz}^{(i)}}{\mathcal{M}_{xz}^{(i)}} \tag{46}$$

$$\ddot{\bar{q}}_{xy}^{(i)} + 2\xi_{xy}^{(i)} \omega_{xy}^{(i)} \dot{\bar{q}}_{xy}^{(i)} + (\omega_{xy}^{(i)})^2 \bar{q}_{xy}^{(i)} = \frac{\bar{Q}_{xy}^{(i)}}{\mathcal{M}_{xy}^{(i)}} \tag{47}$$

$$\ddot{\bar{q}}_{zy}^{(i)} + 2\xi_{zy}^{(i)} \omega_{zy}^{(i)} \dot{\bar{q}}_{zy}^{(i)} + (\omega_{zy}^{(i)})^2 \bar{q}_{zy}^{(i)} = \frac{\bar{Q}_{zy}^{(i)}}{\mathcal{M}_{zy}^{(i)}} \tag{48}$$

$$\begin{aligned} &M_R \ell_R \Big\{ \dot{U}_o \alpha_o + \dot{w} + (L - \ell_{CG}) \ddot{\theta} + \dot{U}_o (\delta_\theta)_o \\ &- \dot{U}_o \sum_i \sigma_{xz}^{(i)}(eG) \bar{q}_{xz}^{(i)} + \sum_i \left[\varphi_{xz}^{(i)}(eG) - \ell_R \sigma_{xz}^{(i)}(eG) \right] \ddot{\bar{q}}_{xz}^{(i)} \Big\} \\ &- \bar{K}_1 \delta_{c\theta} - \bar{K}_2 \dddot{\delta}_\theta + (M_R \ell_R^2 \dot{U}_o - \bar{K}_3) \ddot{\delta}_\theta \\ &+ (\bar{C}_{f_{xz}} - \bar{K}_4) \dot{\delta}_\theta + (M_R \ell_R \dot{U}_o - \bar{K}_5) \delta_\theta = 0 \end{aligned} \tag{49}$$

$$\begin{aligned} &M_R \ell_R \Big\{ \dot{U}_o + \dot{u} + \dot{V}_o + \dot{v} - (L - \ell_{CG}) \ddot{\psi} + \sum_i \left[\varphi_{xy}^{(i)}(eG) - \ell_R \sigma_{xy}^{(i)}(eG) \right] \ddot{\bar{q}}_{xy}^{(i)} \Big\} \\ &- \bar{K}_6 \delta_{c\psi} - \bar{K}_7 \dddot{\delta}_\psi + (M_R \ell_R^2 - \bar{K}_8) \ddot{\delta}_\psi + (\bar{C}_{f_{xy}} - \bar{K}_9) \dot{\delta}_\psi - \bar{K}_{10} \delta_\psi = 0 \end{aligned} \tag{50}$$

$$\begin{aligned} &M_{1\ell} \ddot{Y}_{\ell_{xz}} + 2\xi_\ell \omega_\ell M_{1\ell} \dot{Y}_{\ell_{xz}} + K_\ell Y_{\ell_{xz}} = \\ &M_{1\ell} \left[\dot{w} - \ddot{\theta} X_\ell + \sum_i \varphi_{xz}^{(i)}(X_\ell) \ddot{\bar{q}}_{xz}^{(i)} \right] \end{aligned} \tag{51}$$

$$\begin{aligned} &M_{1\ell} \ddot{Y}_{\ell_{xy}} + 2\xi_\ell \omega_\ell M_{1\ell} \dot{Y}_{\ell_{xy}} + K_\ell Y_{\ell_{xy}} = \\ &M_{1\ell} \left[\dot{v} + \ddot{\psi} X_\ell + \sum_i \varphi_{xy}^{(i)}(X_\ell) \ddot{\bar{q}}_{xy}^{(i)} \right] \end{aligned} \tag{52}$$

Now, in Eqs. (46), (47), and (48) the generalized force terms as developed in Appendix A, are:

$$\bar{Q}_{xz}^{(i)} = \int_0^L w_{xz} \varphi_{xz}^{(i)} \, d\ell \tag{53}$$

$$\bar{Q}_{xy}^{(i)} = \int_0^L w_{xy} \varphi_{xy}^{(i)} \, d\ell \tag{54}$$

$$\bar{Q}_{zy}^{(i)} = \int_0^L \bar{T} \varphi_{zy}^{(i)} \, d\ell \tag{55}$$

w_{xz} and w_{xy} are the external forces per unit length acting perpendicular to the elastic axis in the respective XZ and XY-planes, and $\bar{T}$ is the external moment per unit length about the missile elastic axis. By taking the appropriate component of the forces in Eq. (15) and Eq. (14), Eqs. (53) and (54) become:

$$\begin{aligned}\bar{Q}_{xz}^{(i)} \cong {} & \left\{-T_c\,\delta_\theta + (T_s+T_c)\sum_j \sigma_{xz}^{(j)}(eG)\,\bar{q}_{xz}^{(j)}\right\}\varphi_{xz}^{(i)}(eG) \\ & -\sum_\ell K_\ell Y_{\ell_{xz}}\,\varphi_{xz}^{(i)}(X_\ell) + \sum_\ell\left[M_{c_\ell}\ddot{Y}_{\ell_{xz}} + \frac{M_{I\ell}(T_s+T_c-D)}{M_t}\,Y_{\ell_{xz}}\right]\sigma_{xz}^{(i)}(X_{\ell_B}) \\ & - M_R\left\{\dot{w} + (L-\ell_{CG})\,\ddot{\theta} + \ell_R\,\ddot{\delta}_\theta + \sum_j\left[\varphi_{xz}^{(j)}(eG) - \ell_R\,\sigma_{xz}^{(j)}(eG)\right]\ddot{\bar{q}}_{xz}^{(j)}\right\}\varphi_{xz}^{(i)}(\ell_R) \\ & + \int_0^L \frac{\partial(\Delta Z)}{\partial \ell}\,\varphi_{xz}^{(i)}\,d\ell \end{aligned} \tag{56}$$

$$\begin{aligned}\bar{Q}_{xy}^{(i)} \cong {} & \left\{-T_c\,\delta_\psi + (T_s+T_c)\sum_j \sigma_{xy}^{(j)}(eG)\,\bar{q}_{xy}^{(j)}\right\}\varphi_{xy}^{(i)}(eG) \\ & -\sum_\ell K_\ell Y_{\ell_{xy}}\,\varphi_{xy}^{(i)}(X_\ell) - \sum_\ell\left[M_{c_\ell}\ddot{Y}_{\ell_{xy}} + \frac{M_{I\ell}}{M_t}(T_s+T_c-D)\,Y_{\ell_{xy}}\right]\sigma_{xy}^{(i)}(X_{\ell_B}) \\ & - M_R\left\{\dot{v} - (L-\ell_{CG})\,\ddot{\psi} + \ell_R\,\ddot{\delta}_\psi + \sum_j\left[\varphi_{xy}^{(j)}(eG) - \ell_R\,\sigma_{xy}^{(j)}(eG)\right]\ddot{\bar{q}}_{xy}^{(j)}\right\}\varphi_{xy}^{(i)}(\ell_R) \\ & + \int_0^L \frac{\partial(\Delta Y)}{\partial \ell}\,\varphi_{xy}^{(i)}\,d\ell \end{aligned} \tag{57}$$

And using Eq. (16), Eq. (55) becomes

$$\bar{Q}_{zy}^{(i)} \cong \int_0^L \frac{\partial(\Delta L)}{\partial \ell}\,\varphi_{zy}^{(i)}\,d\ell \tag{58}$$

The integral expressions in Eqs. (56), (57), and (58) will, in nearlv all cases, be evaluated by evaluating the integrand at a large number of stations along the longitudinal axis and summing over the length of the missile.

Equations (7), (8), (9), (10), (11), (12), (46), (47), (48), (49), (50), (51), and (52) are the determinant set of n equations and n unknowns; they are the forward-loop equations of motion with $\delta_{c\theta}$ and $\delta_{c\psi}$ being the input signals to the control engine actuators. These equations represent the total motion of the missile along the trajectory; that is, they contain the steady-state motion as well as the perturbed motions. To investigate the system dynamic stability, however, we need to deal with only the perturbed motions. By setting the disturbance variables equal to zero in the total equations, we can obtain the steady-state equations; then, by subtracting these equations out of the total equations, we obtain the disturbance equations of motion. From this point on, the analysis will deal with only these disturbance equations.

Now, a control system is needed to control and stabilize the missile along a trajectory. The basic elements of a typical automatic missile control system are: the rate and position sensing gyros; the controller (autopilot) elements including amplifiers, integrators, and filters; and a control-engine position actuator. A simplified block diagram of a basic closed-loop system is shown in Figure 5.

Missile control systems are highly complex combinations of many elements; they vary widely in their characteristics, depending on the job they are designed to perform. To complete the closed-loop system representation, equations must be written describing the input-output relations of the various control system elements. Obviously, these equations will be different for each individual control system. The equations presented below are typical of the basic elements described above, but they must be revised for any given analysis to represent the particular control system used.

Typical rate-gyro equations fcr control in each of the three planes are (Ref. 4):

$$\ddot{\theta}_{RG} + 2(\xi_{RG}\,\omega_{RG})_{xz}\,\dot{\theta}_{RG} + (\omega_{RG})^2_{xz}\,\theta_{RG} = (\omega_{RG})^2_{xz}\,K_{R_{xz}}\left[\dot{\theta} + \sum_i \sigma^{(i)}_{xz}(RG)\,\dot{\bar{q}}^{(i)}_{xz}\right] \tag{59}$$

$$\ddot{\psi}_{RG} + 2(\xi_{RG}\,\omega_{RG})_{xy}\,\dot{\psi}_{RG} + (\omega_{RG})^2_{xy}\,\psi_{RG} = (\omega_{RG})^2_{xy}\,K_{R_{xy}}\left[\dot{\psi} + \sum_i \sigma^{(i)}_{xy}(RG)\,\dot{\bar{q}}^{(i)}_{xy}\right] \tag{60}$$

$$\ddot{\phi}_{RG} + 2(\xi_{RG}\,\omega_{RG})_{zy}\,\dot{\phi}_{RG} + (\omega_{RG})^2_{zy}\,\phi_{RG} = (\omega_{RG})^2_{zy}\,K_{R_{zy}}\left[\dot{\phi} + \sum_i \sigma^{(i)}_{zy}(RG)\,\dot{\bar{q}}^{(i)}_{zy}\right] \tag{61}$$

Typical position gyro equations are:

$$\tau_{P_{xz}}\,\dot{\theta}_{PG} + \theta_{PG} = \theta + \sum_i \sigma^{(i)}_{xz}(PG)\,\bar{q}^{(i)}_{xz} \tag{62}$$

$$\tau_{P_{xy}}\,\dot{\psi}_{PG} + \psi_{PG} = \psi + \sum_i \sigma^{(i)}_{xy}(PG)\,\bar{q}^{(i)}_{xy} \tag{63}$$

$$\tau_{P_{zy}}\,\dot{\phi}_{PG} + \phi_{PG} = \phi + \sum_i \sigma^{(i)}_{zy}(PG)\,\bar{q}^{(i)}_{zy} \tag{64}$$

Typical equations describing the controller input-output relations are:

$$\left(\frac{\tau_f}{K_A K_I}\right)_{xz}\ddot{\delta}_{c_\theta} + \left(\frac{1}{K_A K_I}\right)_{xz}\dot{\delta}_{c_\theta} = -\left[\left(\frac{1}{K_I}\right)_{xz} + \tau_{f_{xz}}\right]\dot{\theta}_F + (\theta_c - \theta_F) \tag{65}$$

$$\left(\frac{\tau_f}{K_A K_I}\right)_{xy} \ddot{\delta}_{c_\psi} + \left(\frac{1}{K_A K_I}\right)_{xy} \dot{\delta}_{c_\psi} = -\left[\left(\frac{1}{K_I}\right)_{xy} + \tau_{f_{xy}}\right] \dot{\psi}_F + (\psi_c - \psi_F) \tag{66}$$

$$\left(\frac{\tau_f}{K_A K_I}\right)_{zy} \ddot{\delta}_{c_\phi} + \left(\frac{1}{K_A K_I}\right)_{zy} \dot{\delta}_{c_\phi} = -\left[\left(\frac{1}{K_I}\right)_{zy} + \tau_{f_{zy}}\right] \dot{\phi}_F + (\phi_c - \phi_F) \tag{67}$$

Since we assumed only one control engine was gimballed at the elastic axis, roll control must be accomplished by auxiliary means, such as vernier rockets or reaction jets. Thus δ_{c_ϕ} in Eq. (67) is the autopilot roll-command signal to this auxiliary control, rather than to the gimballed engine. The additional thrust and external moment produced by this intermittently operating auxiliary control is neglected in the missile equations of motion.

The closed-loop system of equations is now complete. Simultaneous solution of the variable coefficient, linear, differential equations, which include the steady-state terms, will yield the complete dynamic response of the missile along the entire trajectory. However, it is usually sufficient to make point-stability analyses involving only the perturbed equations.

Two methods are used primarily to analyze the point stability. One method is to mechanize the equations on an analog or digital computer and generate time-histories of the system variables, or unknowns. An examination of the time-histories will generally reveal whether or not the system is stable at each point analyzed. If one is interested only in the system stability and not the time-history of the oscillations, however, the transfer function formulation is a much more convenient approach. This approach is also a point-stability analysis, since transfer functions can only be formulated for linear, constant-coefficient, differential equations. By the application of Laplace transforms, the differential equations are converted to a like number of simultaneous algebraic equations, to which Cramer's rule can be applied. This results in an input-output relation, which is a ratio of polynomials in the complex variable S, and is termed a transfer function.

After the system of closed-loop equations has been transformed, it can be put in the following matrix form:

$$
\left[A_{jk}\right]\left\{\begin{matrix}
\bar{q}_{xz}^{(i)}(s)\\
\vdots\\
q_{xy}^{(i)}(s)\\
\vdots\\
q_{zy}^{(i)}(s)\\
\vdots\\
u\,(s)\\
v\,(s)\\
w\,(s)\\
\phi\,(s)\\
\theta\,(s)\\
\psi\,(s)\\
Y_{\ell_{xz}}(s)\\
\vdots\\
Y_{\ell_{xy}}(s)\\
\vdots\\
\delta_{\theta}\,(s)\\
\delta_{\psi}\,(s)\\
\delta_{c\theta}\,(s)\\
\delta_{c\psi}\,(s)\\
\delta_{c\phi}\,(s)\\
\theta_{RG}\,(s)\\
\psi_{RG}\,(s)\\
\phi_{RG}\,(s)\\
\theta_{PG}\,(s)\\
\psi_{PG}\,(s)\\
\phi_{PG}\,(s)
\end{matrix}\right\}
= \left[B\right]\left\{v_g(s)\right\} + \left[C\right]\left\{w_g(s)\right\} + \left[D\right]\left\{\theta_c(s)\right\} + \left[E\right]\left\{\psi_c(s)\right\} + \left[F\right]\left\{\phi_c(s)\right\}
$$

where the A_{jk} square matrix and the B, C, D, E, and F diagonal matrices are made up of the coefficients of the variables in each equation, and the coefficients were evaluated at a particular time instant along the trajectory. The condition for the system to be dynamically stable at this point in time is that the polynomial obtained by expanding the characteristic determinant of coefficients has no root with a positive real part. The expansion and factorization can be carried out on a digital computer.

In Appendix E the A_{jk}, B, C, D, E, and F coefficients are tabulated in general form. For simplicity, we assumed that five elastic modes in each plane adequately represent the missile elasticity, and that there are no more than four liquid-propellant tanks.

It should be mentioned that when a control system is being designed, the values for the various gains and time constants of the servo element to insure stability are usually not known. This is a synthesis problem that is generally handled by root-locus and other servo analysis techniques. The conventional method of stability analysis presented in this report can be used only on systems with known or selected values of servo gains, since the gains appear in the equation coefficients.

In applying this analysis to a particular configuration and a given mission, large simplifications can likely be made by neglecting terms known to be small compared to the dominant terms in the equations. For example, on a missile with no large aerodynamic surfaces, aerodynamic terms may possibly be safely omitted, since they would be small compared to the thrust forces and provide negligible coupling (Ref. 7).

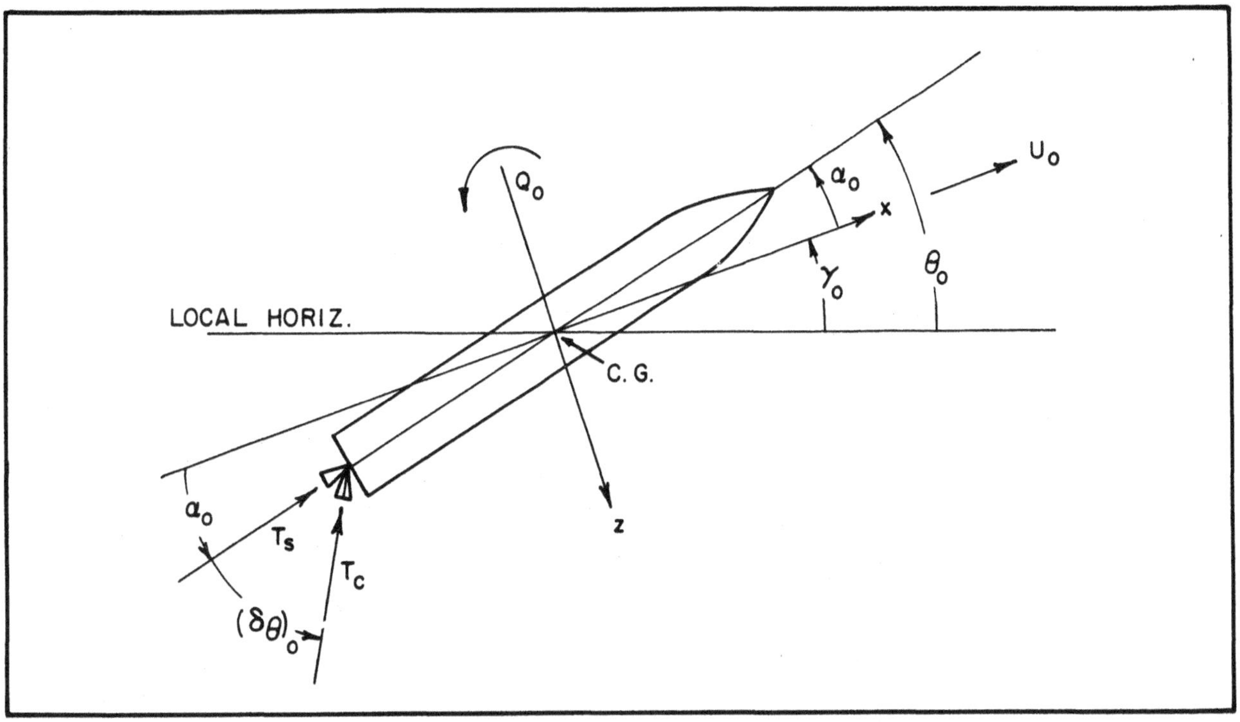

Figure 1. XZ-Plane for Steady Flight

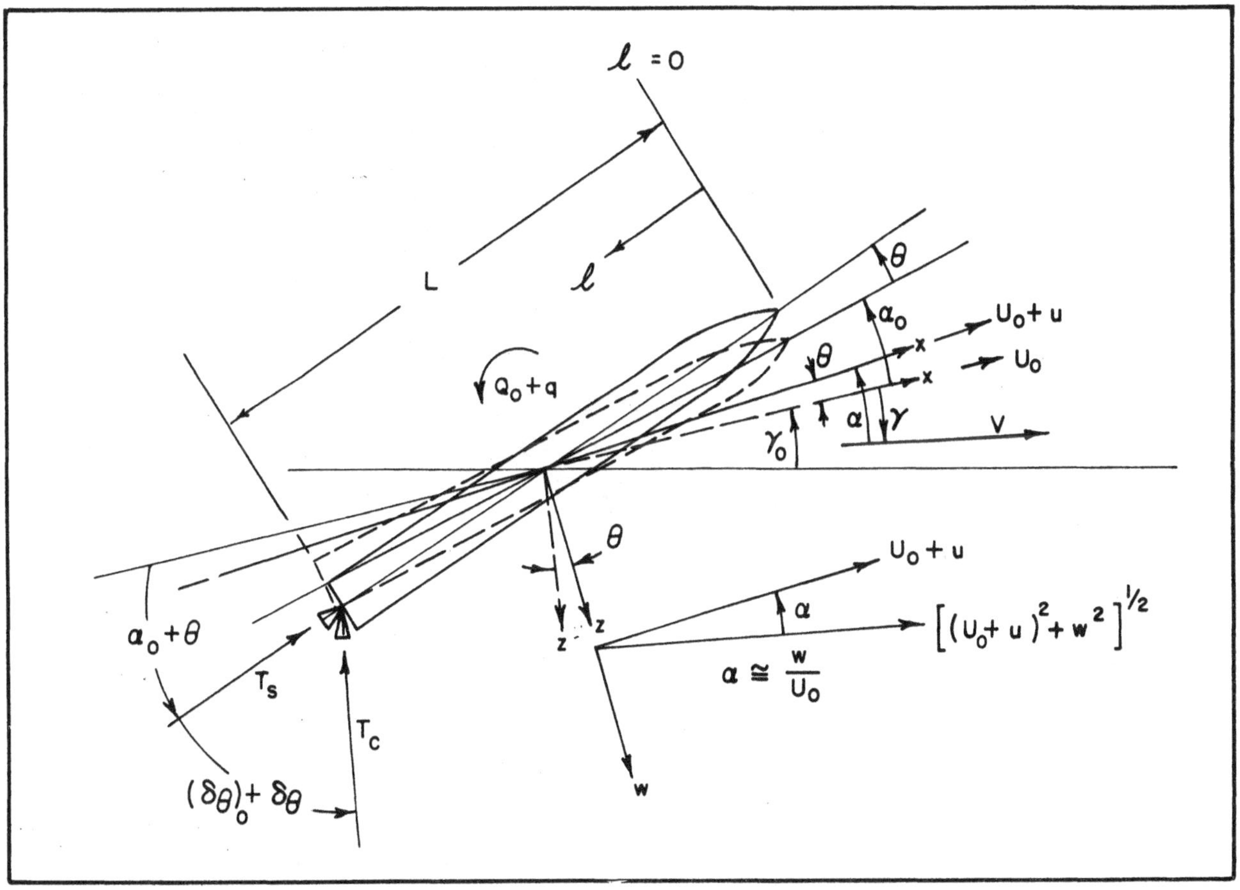

Figure 2. XZ-Plane for Disturbed Flight

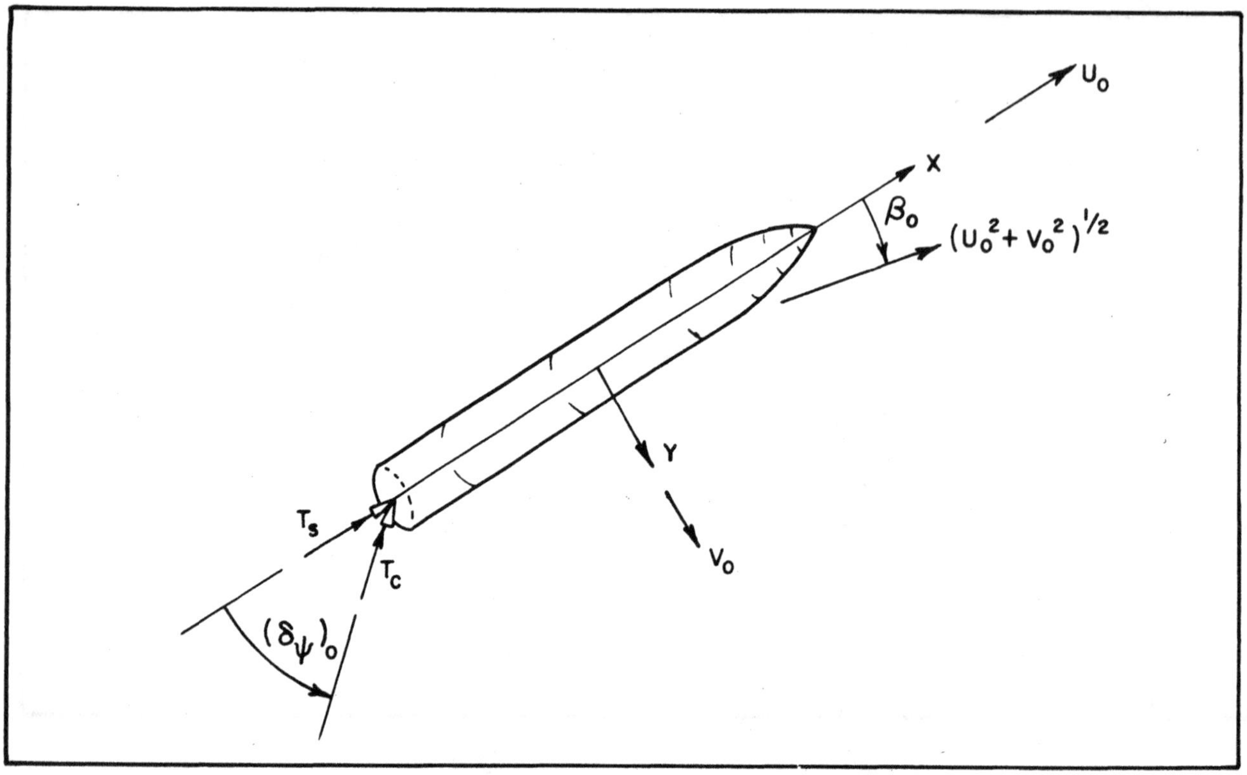

Figure 3. XY-Plane for Steady Flight

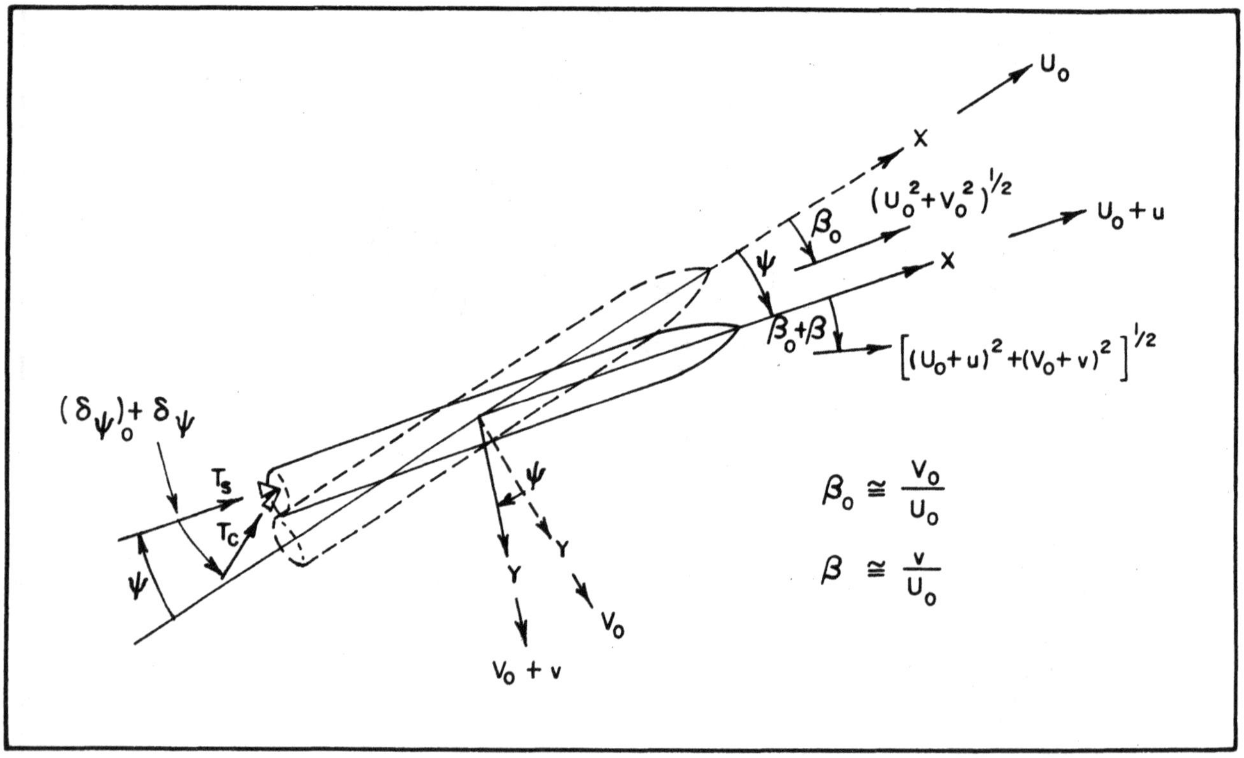

Figure 4. XY-Plane for Disturbed Flight

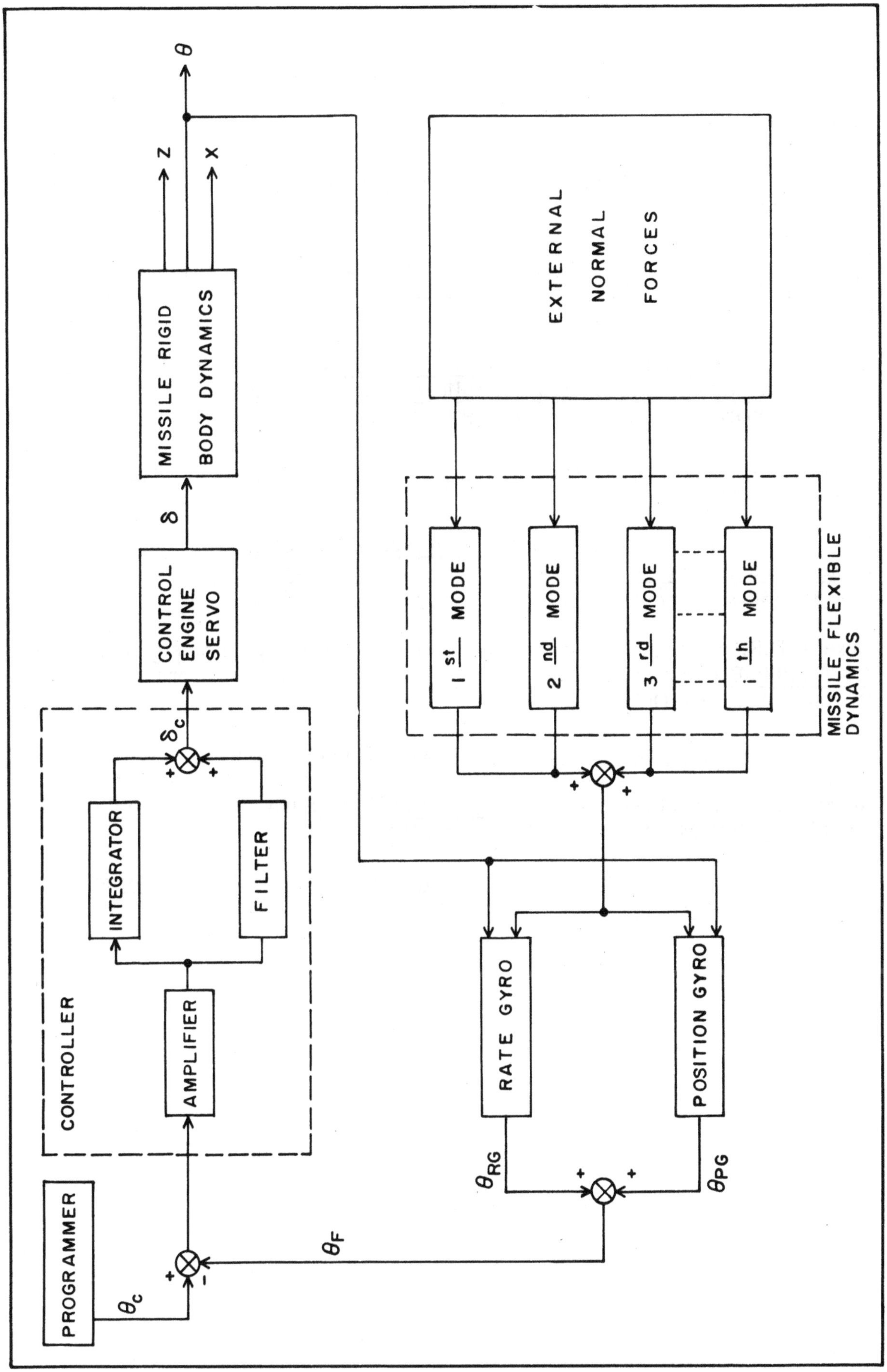

Figure 5. Typical Closed-Loop System

REFERENCES

1. An Introduction to Ballistic Missiles, Vol. III, Astia Document 240179, Mar 1, 1960

2. Ashkenas, I.L., Bates, C.L., and McRuer, D.T., Dynamics of the Airframe, AE-61-4II, Bureau of Aeronautics, Sept, 1952

3. Backus, F.I., Calculation of Closed-Loop Bending Roots for Flexible-Bodied Missiles Using Autopilot Control By a Method Adaptable to Manual Calculation, Convair Report AE 60-0557, Oct 5, 1960

4. Backus, F.I., Describing Functions for Nonlinear Electrohydraulic Gimballed Rocket-Engine Position Servos with Application to Closed-Loop Control Systems, Convair Report AE 60-0287, June 10, 1960.

5. Kachigan, K., The General Theory and Analysis of a Flexible Bodied Missile with Autopilot Control, Convair Report ZU-7-048, Nov 15, 1955.

6. Kelly, C.P., and Lukens, D.R., (U) The Influence of Propellant Sloshing Upon the Dynamic Stability of the XSM-65A and the XSM-65C Missiles During First Stage, Convair Report ZU-7-054-TN, Feb 2, 1956, CONFIDENTIAL.

7. Lukens, D.R., Schmitt, A.F., and Broucek, G.T., Approximate Transfer Functions for Flexible Booster and Autopilot Analysis, WADD TR 61-93, Aeronautical Systems Division, W-PAFB, Ohio. April 1961.

8. Myklestad, N.O., Vibration Analysis, McGraw Hill Book Co., Inc., New York, 1944.

9. Rosenbaum, R., and Scanlan, R.H., Introduction to the Study of Aircraft Vibration and Flutter, The Macmillan Co., New York, 1951.

10. Schmitt, A.F., Forced Oscillations of a Fluid in a Cylindrical Tank Undergoing Both Translation and Rotation, Convair Report ZU-7-069, Oct 16, 1956.

APPENDIX A

STRUCTURAL DYNAMICS

APPENDIX A

STRUCTURAL DYNAMICS

In this appendix, the generalized equations for forced vibration of a free-free elastic missile are developed; solutions of these equations yield the time-dependent elastic deflections of the missile. Three sets of modal equations are derived corresponding to bending vibrations in the XZ-plane and XY-plane and torsional vibration parallel to the ZY-plane (assuming small α_0).

Figure 6 is a schematic diagram of the elastic missile in the XZ-plane. The general equation for forced vibration is (based on simple beam theory):

$$\frac{\partial^2}{\partial \ell^2}\left[E_{xz}(\ell)\, I_{xz}(\ell)\, \frac{\partial^2 \bar{u}_{xz}}{\partial \ell^2}\right] = -m(\ell)\, \frac{\partial^2 \bar{u}_{xz}}{\partial t^2} + w_{xz}(\ell, t)\,, \qquad (1A)$$

where $w_{xz}(\ell, t)$ is the external load distribution over the missile perpendicular to the elastic axis.

The solution of Eq. (1A) is obtained by separation of variables. The deflection $\bar{u}_{xz}$ from the undeformed elastic axis is taken as the product of two functions, one of which is a function of time, and the other a function of the location coordinate ℓ.

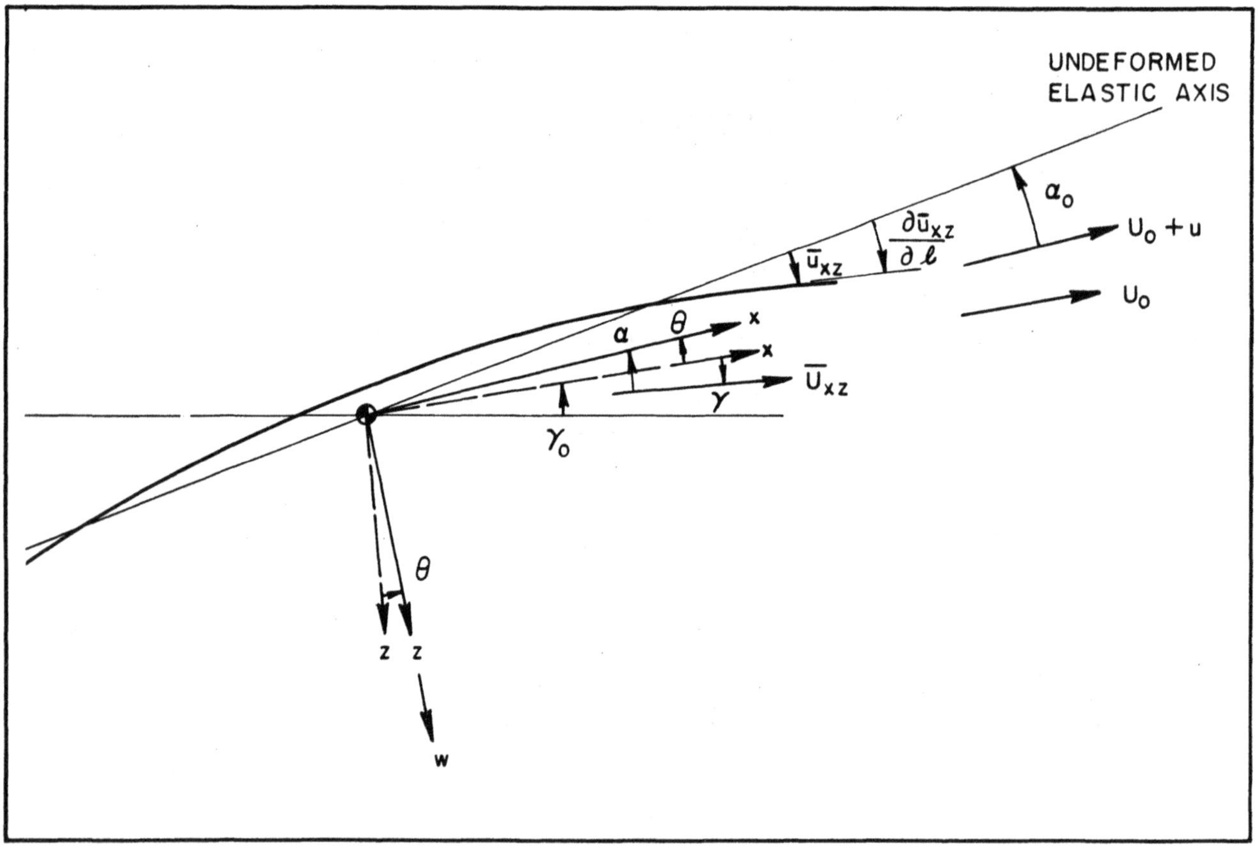

Figure 6. Elastic Missile in Disturbed Flight in XZ-Plane

$$\bar{u}_{xz} = F_{xz}(t)\, \phi_{xz}(\ell) \tag{2A}$$

Substituting Eq. (2A) into the homogeneous equation obtained from Eq. (1A) and separating variables, yields:

$$\frac{d^2 F}{dt^2} + (\omega^{(i)})^2 F = 0 \tag{3A}$$

$$\frac{d^2}{d\ell^2}\left[E(\ell)\, I(\ell)\, \frac{d^2\phi}{d\ell^2}\right] - m(\ell)\,(\omega^{(i)})^2\, \phi = 0 \tag{4A}$$

$(\omega^{(i)})^2$ is a constant which is undertermined as yet. Eq. (3A) has a sinusoidal solution of the form below, which indicates that $\omega^{(i)}$ has the units of rad/sec

$$F_{xz}(t) = B_1 \sin \omega^{(i)} t + B_2 \cos \omega^{(i)} t, \tag{5A}$$

It can be shown that Eq. (4A) admits an infinite set of eigenvalues $\omega^{(i)}$; and for each $\omega^{(i)}$ there is a corresponding eigenfunction $\phi_{xz}^{(i)}(\ell)$, which possesses the following property of orthogonality:

$$\int_0^L m(\ell)\, \phi_{xz}^{(i)}(\ell)\, \phi_{xz}^{(j)}(\ell)\, d\ell = a_{ij}\, \delta_{ij} \quad , \tag{6A}$$

where $\delta_{ij} = 1$ if $i = j$, and $\delta_{ij} = 0$ if $i \neq j$.
If the eigenvalues form an infinite sequence of distinct numbers and the orthogonality condition holds, then any arbitrary, piece-wise, continuous function can be represented by a series of $\phi_{xz}^{(i)}(\ell)$ in the open interval $0 < \ell < L$.

The homogeneous differential equation of motion of the free-free missile is a linear equation such that each solution of Eqs. (3A) and (4A) is a solution of the differential equation; and, furthermore, the sum of these solutions is also a solution. The most general solution for the deflection function can be written as:

$$\bar{u}_{xz}(\ell, t) = \sum_{i=1}^{\infty} F_{xz}^{(i)}(t)\, \phi_{xz}^{(i)}(\ell) \tag{7A}$$

It will be assumed that the $\phi_{xz}^{(i)}(\ell)$ and $\omega^{(i)}$ have been previously calculated by one of a variety of methods (Ref. 9).

Let $\varphi_{xz}^{(i)}(\ell)$ represent the orthogonal functions which were normalized at some point on the missile, such as the gimbal point. In this case, the relation between $\phi_{xz}^{(i)}(\ell)$ and $\varphi_{xz}^{(i)}(\ell)$ would be:

$$\varphi_{xz}^{(i)}(\ell) = \frac{\phi_{xz}^{(i)}(\ell)}{\phi_{xz}^{(i)}(eG)} \tag{8A}$$

Now, if we let

$$\bar{q}_{xz}^{(i)}(t) = \phi_{xz}^{(i)}(eG)\; F_{xz}^{(i)}(t) \quad , \tag{9A}$$

it then follows from Eq. (7A) that:

$$\bar{u}_{xz}(\ell, t) = \sum_{i=1}^{\infty} \bar{q}_{xz}^{(i)}(t)\; \varphi_{xz}^{(i)}(\ell) \tag{10A}$$

Now, assume that the external force distribution can be represented by a series similar to $\bar{u}_{xz}(\ell, t)$. In particular, assume

$$w_{xz}(\ell, t) = \sum_{i=1}^{\infty} A^{(i)}(t)\; m(\ell)\; \varphi_{xz}^{(i)}(\ell) \tag{11A}$$

where $A^{(i)}(t)$ are the undetermined functions. Multiplying both sides of Eq. (11A) by $\varphi_{xz}^{(j)}(\ell)$ and integrating over the length of the missile results in:

$$\int_0^L \varphi_{xz}^{(j)}(\ell)\; w_{xz}(\ell, t)\, d\ell = \sum_{i=1}^{\infty} A^{(i)}(t) \int_0^L m(\ell)\; \varphi_{xz}^{(i)}(\ell)\; \varphi_{xz}^{(j)}(\ell)\; d\ell \tag{12A}$$

Due to the orthogonality relationship of the eigenfunctions, all product terms vanish except when $i = j$; therefore,

$$A^{(i)}(t) = \frac{\int_0^L \varphi_{xz}^{(i)}(\ell)\; w_{xz}(\ell, t)\; d\ell}{\int_0^L m(\ell) \left[\varphi_{xz}^{(i)}(\ell) \right]^2 d\ell} \tag{13A}$$

The numerator of Eq. (13A) is designated as the generalized force

$$\bar{Q}_{xz}^{(i)}(t) = \int_0^L w_{xz}(\ell, t)\; \varphi_{xz}^{(i)}(\ell)\; d\ell \tag{14A}$$

while the denominator is defined as the generalized mass.

$$\mathcal{M}_{xz}^{(i)} = \int_0^L m(\ell) \left[\varphi_{xz}^{(i)}(\ell) \right]^2 d\ell \tag{15A}$$

Therefore,

$$A^{(i)}(t) = \frac{\bar{Q}_{xz}^{(i)}(t)}{\mathcal{M}_{xz}^{(i)}} \tag{16A}$$

By substituting Eq. (16A) into Eq. (11A), the loading function becomes:

$$w_{xz}(\ell, t) = \sum_{i=1}^{\infty} \frac{\bar{Q}_{xz}^{(i)}(t)}{\mathcal{M}_{xz}^{(i)}} \, m(\ell) \, \varphi_{xz}^{(i)}(\ell) \tag{17A}$$

By substituting Eqs. (10A) and (17A) into Eq. (1A) it is seen that:

$$\frac{1}{m(\ell)\,\varphi_{xz}^{(i)}(\ell)} \frac{\partial^2}{\partial \ell^2}\left[E_{xz}(\ell)\, I_{xz}(\ell)\, \bar{q}_{xz}^{(i)}(t) \frac{\partial^2 \varphi_{xz}^{(i)}(\ell)}{\partial \ell^2}\right] + \ddot{\bar{q}}_{xz}^{(i)} = \frac{\bar{Q}_{xz}^{(i)}(t)}{\mathcal{M}_{xz}^{(i)}} \tag{18A}$$

Now, from Eq. (4A):

$$(\omega^{(i)})^2 = \frac{1}{m(\ell)\,\varphi_{xz}^{(i)}(\ell)} \frac{\partial^2}{\partial \ell^2}\left[E_{xz}(\ell)\, I_{xz}(\ell) \frac{\partial^2 \varphi_{xz}^{(i)}(\ell)}{\partial \ell^2}\right] \tag{19A}$$

Eq. (18A) then becomes:

$$\ddot{\bar{q}}_{xz}^{(i)} + (\omega^{(i)})^2 \, \bar{q}_{xz}^{(i)} = \frac{\bar{Q}_{xz}^{(i)}(t)}{\mathcal{M}_{xz}^{(i)}} \qquad (i = 1, 2, 3, \cdots) \tag{20A}$$

The missile will possess some dissipative forces which provide system damping. This dissipative energy is usually small in comparison to the elastic and/or kinetic energy, such that the effects of the lower eigenvalues and eigenfunctions are negligible. Therefore, a small equivalent viscous damping term will be included in Eq. (20A) as follows:

$$\ddot{\bar{q}}_{xz}^{(i)}(t) + 2\xi_{xz}^{(i)} \, \omega_{xz}^{(i)} \, \dot{\bar{q}}_{xz}^{(i)}(t) + (\omega^{(i)})_{xz}^2 \, \bar{q}_{xz}^{(i)}(t) = \frac{\bar{Q}_{xz}^{(i)}(t)}{\mathcal{M}_{xz}^{(i)}} \tag{21A}$$

The value of the damping ratio $\xi^{(i)}$ should be determined by experiment. The use of an equivalent viscous term in place of a combination of structural damping, coulomb damping, and true viscous effects is acceptable because of its small magnitude and because of the near-harmonic nature of the motions being investigated.

The solution of Eq. (21A) depends on the determination of the generalized mass $\mathcal{M}_{xz}^{(i)}$ and the generalized forces $\bar{Q}_{xz}^{(i)}(t)$. The generalized mass, as defined by Eq. (15A), is dependent only upon the mass distribution of the missile and the free-free eigenfunctions. The generalized forces, $\bar{Q}_{xz}^{(i)}(t)$, however, are functions of the external loading and the free-free eigenfunctions.

Once these $\bar{Q}_{xz}^{(i)}(t)$ and $\mathcal{M}_{xz}^{(i)}$ functions are formed, $\bar{q}_{xz}^{(i)}(t)$ is determined for each eigenvalue from Eq. (21A). The total deflection, $\bar{u}_{xz}(\ell, t)$, is then obtained from Eq. (10A).

A similar derivation can be made for lateral modal vibrations, which, for small α_0, can be treated as motion in the XY-plane (see Fig. 7). Thus:

$$\ddot{\bar{q}}_{xy}^{(i)} + 2\xi_{xy}^{(i)}\omega_{xy}^{(i)}\dot{\bar{q}}_{xy}^{(i)} + (\omega^{(i)})_{xy}^{2}\bar{q}_{xy}^{(i)} = \frac{\bar{Q}_{xy}^{(i)}}{\mathcal{M}_{xy}^{(i)}} \tag{22A}$$

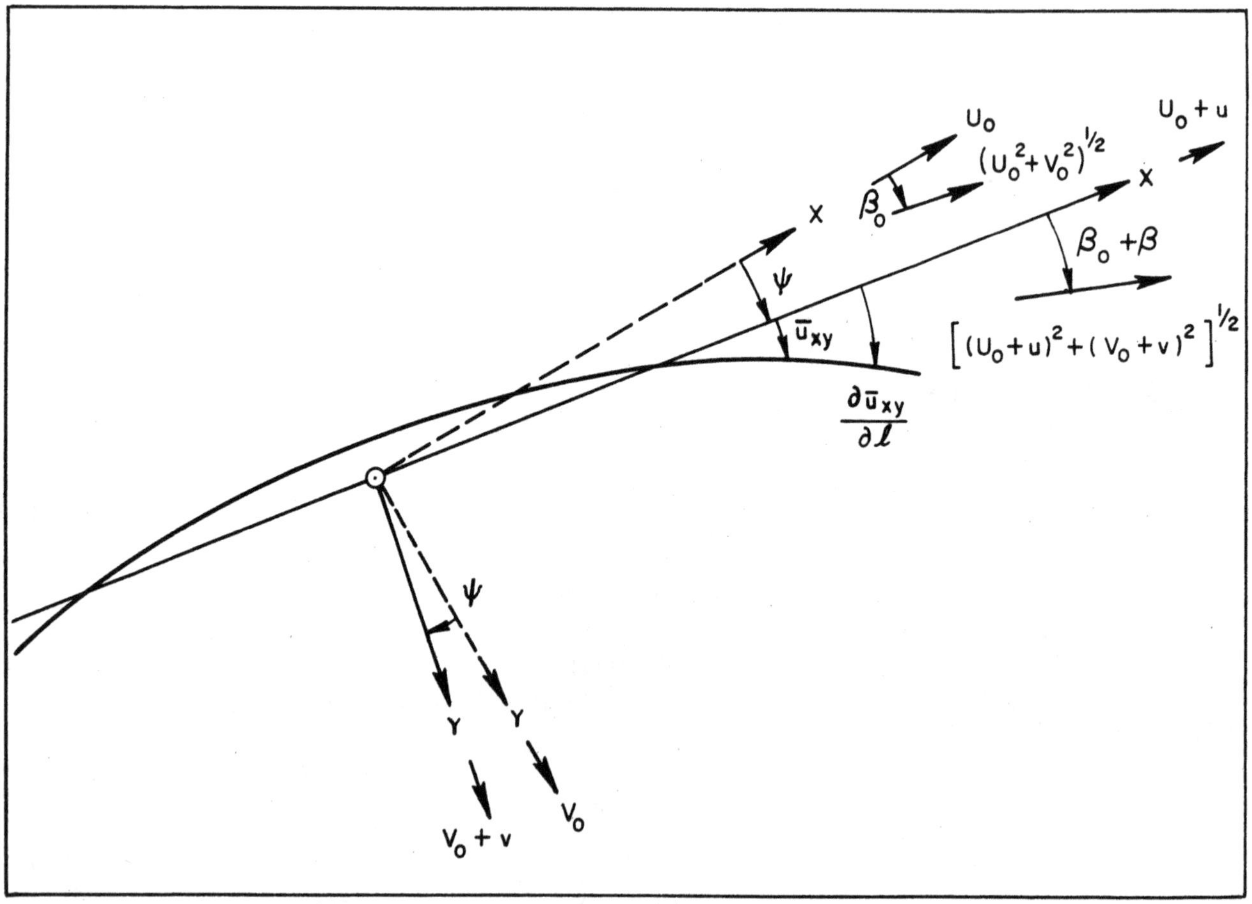

Figure 7. Elastic Missile in Disturbed Flight in XY-Plane

A similar derivation can also be made for torsional modal vibration about the longitudinal axis by starting with the general equation for forced torsional vibration:

$$\frac{\partial}{\partial \ell}\left[G(\ell)\,J(\ell)\,\frac{\partial^2 \bar{u}_{zy}}{\partial \ell}\right] = \bar{I}(\ell)\,\frac{\partial^2 \bar{u}_{zy}}{\partial t^2} - \bar{T}(\ell, t) \tag{23A}$$

where $\bar{u}_{zy}$ is the elastic rotation angle about the longitudinal axis, which, for small α_0 can be considered as rotation about the X-axis and parallel to the ZY-plane. $\bar{I}$ is the mass moment of inertia per unit length about the longitudinal axis, and $\bar{T}$ is the externally applied torque. Again assuming:

$$\bar{u}_{zy} = \sum_i F_{zy}^{(i)}(t)\,\phi_{zy}^{(i)}(\ell) \tag{24A}$$

and proceeding as before, it can be shown that

$$\ddot{\bar{q}}_{zy}^{(i)} + 2\,\xi_{zy}^{(i)}\,\omega_{zy}^{(i)}\,\dot{\bar{q}}_{zy}^{(i)} + (\omega^{(i)})^2_{zy}\,\bar{q}_{zy}^{(i)} = \frac{\bar{Q}_{zy}^{(i)}}{\mathcal{M}_{zy}^{(i)}} \tag{25A}$$

where:

$$\bar{u}_{zy} = \sum_i \bar{q}_{zy}^{(i)}\,\varphi_{zy}^{(i)}(\ell) \quad , \tag{26A}$$

$$\bar{Q}_{zy}^{(i)} = \int_0^L \bar{T}(\ell,t)\,\varphi_{zy}^{(i)}(\ell)\,d\ell \quad , \tag{27A}$$

and

$$\mathcal{M}_{zy}^{(i)} = \int_0^L \bar{I}(\ell)\left[\varphi_{zy}^{(i)}(\ell)\right]^2 d\ell. \tag{28A}$$

APPENDIX B

ENGINE INERTIA

APPENDIX B

ENGINE INERTIA

Figure 8 shows a schematic diagram of a control engine relative to its position in the elastic missile at the disturbed flight condition in the XZ-plane. Because of the complexity of the coordinate system, Lagrangean mechanics are used to determine the engine inertia forces acting on the missile in the XZ-plane:

$$\frac{d}{dt}\left(\frac{\partial \bar{L}}{\partial \dot{x}}\right) - \frac{\partial \bar{L}}{\partial x} = -F_x \tag{1B}$$

and

$$\frac{d}{dt}\left(\frac{\partial \bar{L}}{\partial \dot{z}}\right) - \frac{\partial \bar{L}}{\partial z} = -F_z\ , \tag{2B}$$

where $\bar{L} = \bar{T} - \bar{V}$ = kinetic potential,

$\bar{T}$ = engine kinetic energy, and

$\bar{V}$ = engine potential energy.

$\bar{V}$ is due to gravity; since engine gravity terms have been included in the complete equations of missile motion, $\bar{V} = 0$ for this formulation.

$-F_x$ and $-F_z$ are the forces required to produce the various accelerations on the engine; F_x and F_z, however, are the inertia forces acting on the missile due to these accelerations.

The location of the engine center of gravity with respect to the inertial axes is given by

$$x' = x_0 - \ell_R \cos\left(\alpha_0 + (\delta_\theta)_0 + \delta_\theta + \theta + \frac{\partial \bar{u}_{xz}(eG)}{\partial \ell}\right) \tag{3B}$$

$$z' = z_0 + \ell_R \sin\left(\alpha_0 + (\delta_\theta)_0 + \delta_\theta + \theta + \frac{\partial \bar{u}_{xz}(eG)}{\partial \ell}\right) \tag{4B}$$

The velocity components of the engine C.G. are

$$\dot{x}' = \dot{x}_0 + \ell_R\left(\dot{\theta} + \dot{\delta}_\theta + \frac{\partial^2 \bar{u}_{xz}(eG)}{\partial \ell\, \partial t}\right)\sin\left(\alpha_0 + (\delta_\theta)_0 + \delta_\theta + \theta + \frac{\partial \bar{u}_{xz}(eG)}{\partial \ell}\right)$$

or, neglecting small products,

$$\dot{x}' \cong \dot{x}_0$$

$$\dot{z}' = \dot{z}_0 + \ell_R\left(\dot{\theta} + \dot{\delta}_\theta + \frac{\partial^2 \bar{u}_{xz}(eG)}{\partial \ell \, \partial t}\right)\cos\left(\alpha_0 + (\delta_\theta)_0 + \delta_\theta + \theta + \frac{\partial \bar{u}_{xz}(eG)}{\partial \ell}\right)$$

$$\dot{z}' \cong \dot{z}_0 + \ell_R\left(\dot{\theta} + \dot{\delta}_\theta + \frac{\partial^2 \bar{u}_{xz}(eG)}{\partial \ell \, \partial t}\right)$$

Now, $\dot{x}_0 \cong \dot{x}$, and $z_0 = z'_{CG} + \bar{u}_{xz}(eG)\cos(\alpha_0 + \theta) + (\alpha_0 + \theta)\,\ell_e$
or

$$\dot{z}_0 \cong \dot{z} + \dot{\bar{u}}_{xz}(eG)\cos(\alpha_0 + \theta) - \dot{\theta}\,\bar{u}_{xz}(eG)\sin(\alpha_0 + \theta) + \dot{\theta}\,\ell_e$$

$$\dot{z}_0 \cong \dot{z} + \dot{\bar{u}}_{xz}(eG) + \dot{\theta}\,\ell_e$$

The engine kinetic energy is given by:

$$\bar{T} = \bar{L} = \frac{M_R}{2}\left\{(\dot{x}')^2 + (\dot{z}')^2\right\}$$

$$\bar{T} = \bar{L} \cong \frac{M_R}{2}\left\{\dot{x}^2 + \left[\dot{z} + \dot{\bar{u}}_{xz}(eG) + \dot{\theta}\ell_e + \ell_R\left(\dot{\theta} + \dot{\delta}_\theta + \frac{\partial^2 \bar{u}_{xz}(eG)}{\partial \ell \, \partial t}\right)\right]^2\right\} \qquad (5B)$$

The following derivatives are obtained from Eq. (5B)

$$\frac{\partial \bar{L}}{\partial \dot{x}} \cong M_R \dot{x} \qquad (6B)$$

$$\frac{\partial}{\partial t}\left(\frac{\partial \bar{L}}{\partial \dot{x}}\right) \cong M_R \ddot{x} \qquad (7B)$$

$$\frac{\partial \bar{L}}{\partial x} \cong 0 \qquad (8B)$$

$$\frac{\partial \bar{L}}{\partial \dot{z}} = M_R\left[\dot{z} + \dot{\bar{u}}_{xz}(eG) + \dot{\theta}\ell_e + \ell_R\left(\dot{\theta} + \dot{\delta}_\theta + \frac{\partial^2 \bar{u}_{xz}(eG)}{\partial \ell \, \partial t}\right)\right] \qquad (9B)$$

$$\frac{\partial}{\partial t}\left(\frac{\partial \bar{L}}{\partial \dot{z}}\right) = M_R\left[\ddot{z} + \ddot{\bar{u}}_{xz}(eG) + \ddot{\theta}\ell_e + \ell_R\left(\ddot{\theta} + \ddot{\delta}_\theta + \frac{\partial^3 \bar{u}_{xz}(eG)}{\partial \ell \, \partial t^2}\right)\right] \qquad (10B)$$

$$\frac{\partial \bar{L}}{\partial z} = 0 \qquad (11B)$$

Now, from the Lagrangean equations

$$F_x = -M_R \ddot{X} \tag{12B}$$

$$F_z = -M_R\left[\ddot{Z} + \ddot{\bar{u}}_{xz}(eG) + \ddot{\theta}\ell_e + \ell_R\left(\ddot{\theta} + \ddot{\delta}_\theta + \frac{\partial^3\bar{u}_{xz}(eG)}{\partial \ell\ \partial t^2}\right)\right] \tag{13B}$$

The moment about the missile C.G. due to these engine inertia forces is given by:

$$M_{ei} = \left[F_x \sin\alpha_o + F_z \cos\alpha_o\right]\left[\ell_e + \ell_R \cos\left((\delta_\theta)_o + \delta_\theta + \frac{\partial \bar{u}_{xz}(eG)}{\partial \ell}\right)\right] \tag{14B}$$

For small disturbances and since $\ddot{X} \cong \dot{U}_o + \dot{u}$, $\ddot{Z} \cong \dot{w}$

$$M_{ei} = -(\ell_e + \ell_R)\left\{M_R \dot{U}_o \alpha_o + M_R\left[\dot{w} + \ddot{\bar{u}}_{xz}(eG) + \ddot{\theta}\ell_e + \ell_R\left(\ddot{\theta} + \ddot{\delta}_\theta + \frac{\partial^3\bar{u}_{xz}(eG)}{\partial \ell\ \partial t^2}\right)\right]\right\} \tag{15B}$$

Now, to derive the equation of motion of the engine about the gimbal, moments are summed about that point (see Figure 8):

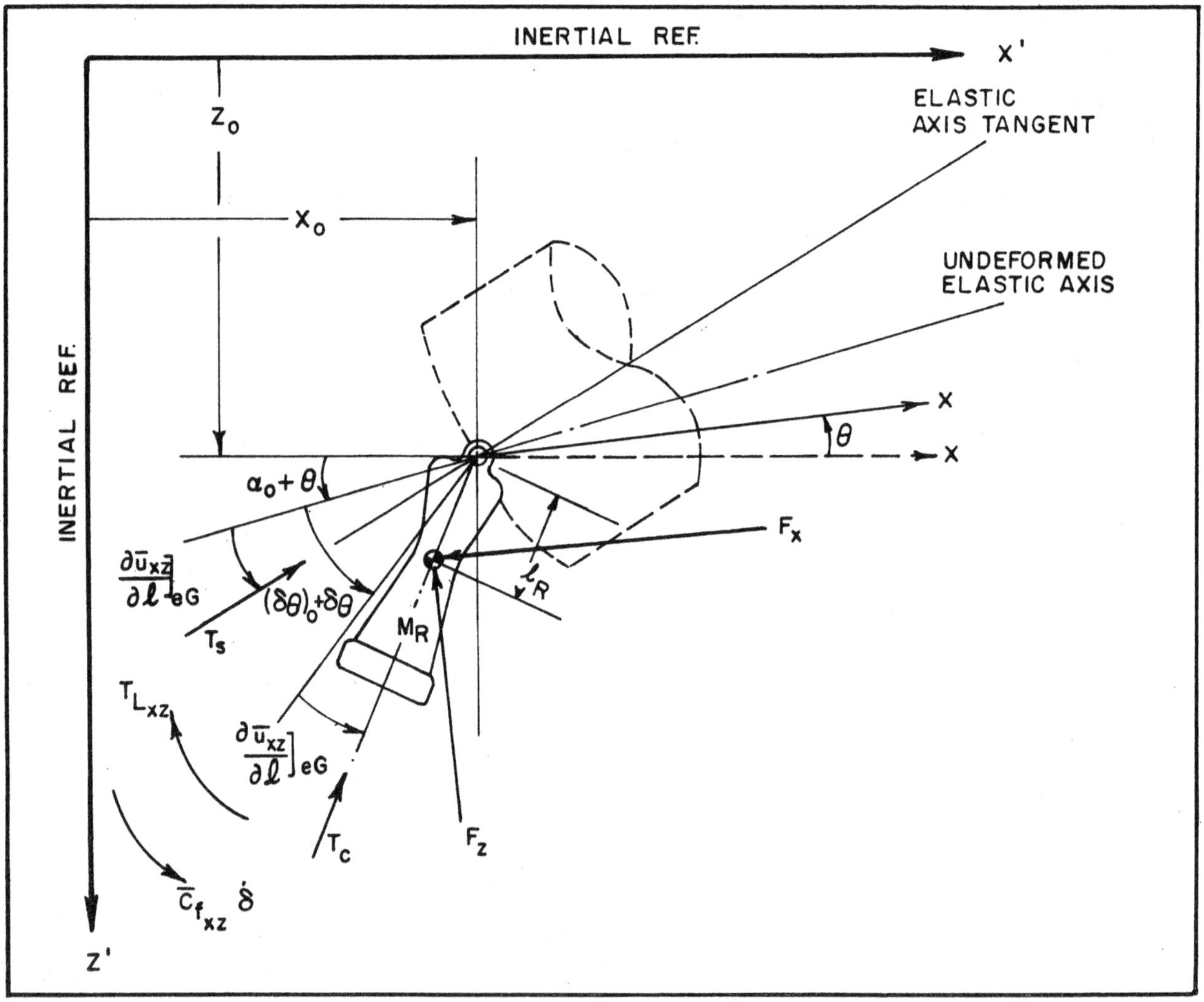

Figure 8. Control Engine in Disturbed Flight in XZ-Plane

$$\ell_R \left[F_z \cos\left(\alpha_0 + (\delta_\theta)_0 + \delta_\theta + \frac{\partial \bar{u}_{xz}(eG)}{\partial \ell}\right) + F_x \sin\left(\alpha_0 + (\delta_\theta)_0 + \delta_\theta + \frac{\partial \bar{u}_{xz}(eG)}{\partial \ell}\right)\right] + T_{L_{xz}} - \bar{C}_{f_{xz}} \dot{\delta}_\theta = 0 \qquad (16B)$$

where $T_{L_{xz}}$ is the torque generated by the servo actuation device in the XZ-plane, and $\bar{C}_{f_{xz}} \dot{\delta}_\theta$ is the gimbal friction torque in the XZ-plane.

Equation (16B), through rearranging and substitution, becomes:

$$T_{L_{xz}} \cong \bar{C}_{f_{xz}} \dot{\delta}_\theta + M_R \ell_R \left\{ \dot{U}_0 \alpha_0 + \dot{U}_0 \left((\delta_\theta)_0 + \delta_\theta + \frac{\partial \bar{u}_{xz}(eG)}{\partial \ell} \right) + \dot{w} + \ddot{\bar{u}}_{xz}(eG) + \ddot{\theta}\ell_e + \ell_R \left(\ddot{\theta} + \ddot{\delta}_\theta + \frac{\partial^3 \bar{u}_{xz}(eG)}{\partial \ell \, \partial t^2} \right) \right\} \qquad (17B)$$

The following relations were obtained from Appendix A and the arbitrary definition $\sigma(\ell) = -\frac{\partial \varphi(\ell)}{\partial \ell}$

$$\bar{u}_{xz}(eG) = \sum_i \varphi_{xz}^{(i)}(eG)\, \bar{q}_{xz}^{(i)} \qquad \frac{\partial \bar{u}_{xz}(eG)}{\partial \ell} = -\sum_i \sigma_{xz}^{(i)}(eG)\, \bar{q}_{xz}^{(i)}$$

$$\dot{\bar{u}}_{xz}(eG) = \sum_i \varphi_{xz}^{(i)}(eG)\, \dot{\bar{q}}_{xz}^{(i)} \qquad \frac{\partial^2 \bar{u}_{xz}(eG)}{\partial \ell \, \partial t} = -\sum_i \sigma_{xz}^{(i)}(eG)\, \dot{\bar{q}}_{xz}^{(i)}$$

$$\ddot{\bar{u}}_{xz}(eG) = \sum_i \varphi_{xz}^{(i)}(eG)\, \ddot{\bar{q}}_{xz}^{(i)} \qquad \frac{\partial^3 \bar{u}_{xz}(eG)}{\partial \ell \, \partial t^2} = -\sum_i \sigma_{xz}^{(i)}(eG)\, \ddot{\bar{q}}_{xz}^{(i)}$$

Using the above relations Eqs. (12B), (13B), (15B), and (17B) become

$$F_x = -M_R \left[\dot{U}_0 + \dot{u} \right] \qquad (18B)$$

$$F_z = -M_R \left\{ \dot{w} + (\ell_R + \ell_e)\ddot{\theta} + \ell_R \ddot{\delta}_\theta + \sum_i \left[\varphi_{xz}^{(i)}(eG) - \ell_R \sigma_{xz}^{(i)}(eG) \right] \ddot{\bar{q}}_{xz}^{(i)} \right\} \qquad (19B)$$

$$M_{ei} = -M_R (\ell_R + \ell_e) \left\{ \dot{U}_0 \alpha_0 + \left[\dot{w} + (\ell_R + \ell_e)\ddot{\theta} + \ell_R \ddot{\delta}_\theta + \sum_i \left[\varphi_{xz}^{(i)}(eG) - \ell_R \sigma_{xz}^{(i)}(eG) \right] \ddot{\bar{q}}_{xz}^{(i)} \right] \right\} \qquad (20B)$$

$$T_{L_{xz}} = M_R \ell_R \left\{ \dot{w} + (\ell_R + \ell_e)\ddot{\theta} + \sum_i \left[\varphi_{xz}^{(i)}(eG) - \ell_R \sigma_{xz}^{(i)}(eG) \right] \ddot{\bar{q}}_{xz}^{(i)} + \dot{U}_0 \left(\alpha_0 + (\delta_\theta)_0 + \delta_\theta + \ell_R \ddot{\delta}_\theta - \sum_i \sigma_{xz}^{(i)}(eG)\, \bar{q}_{xz}^{(i)} \right) \right\} + \bar{C}_{f_{xz}} \dot{\delta}_\theta \qquad (21B)$$

Figure 9 is a schematic of the control engine relative to its position in the elastic missile at the disturbed flight condition in the XY-plane. Lagrange's equations are again used to determine the engine inertia forces acting on the missile in the XY-plane.

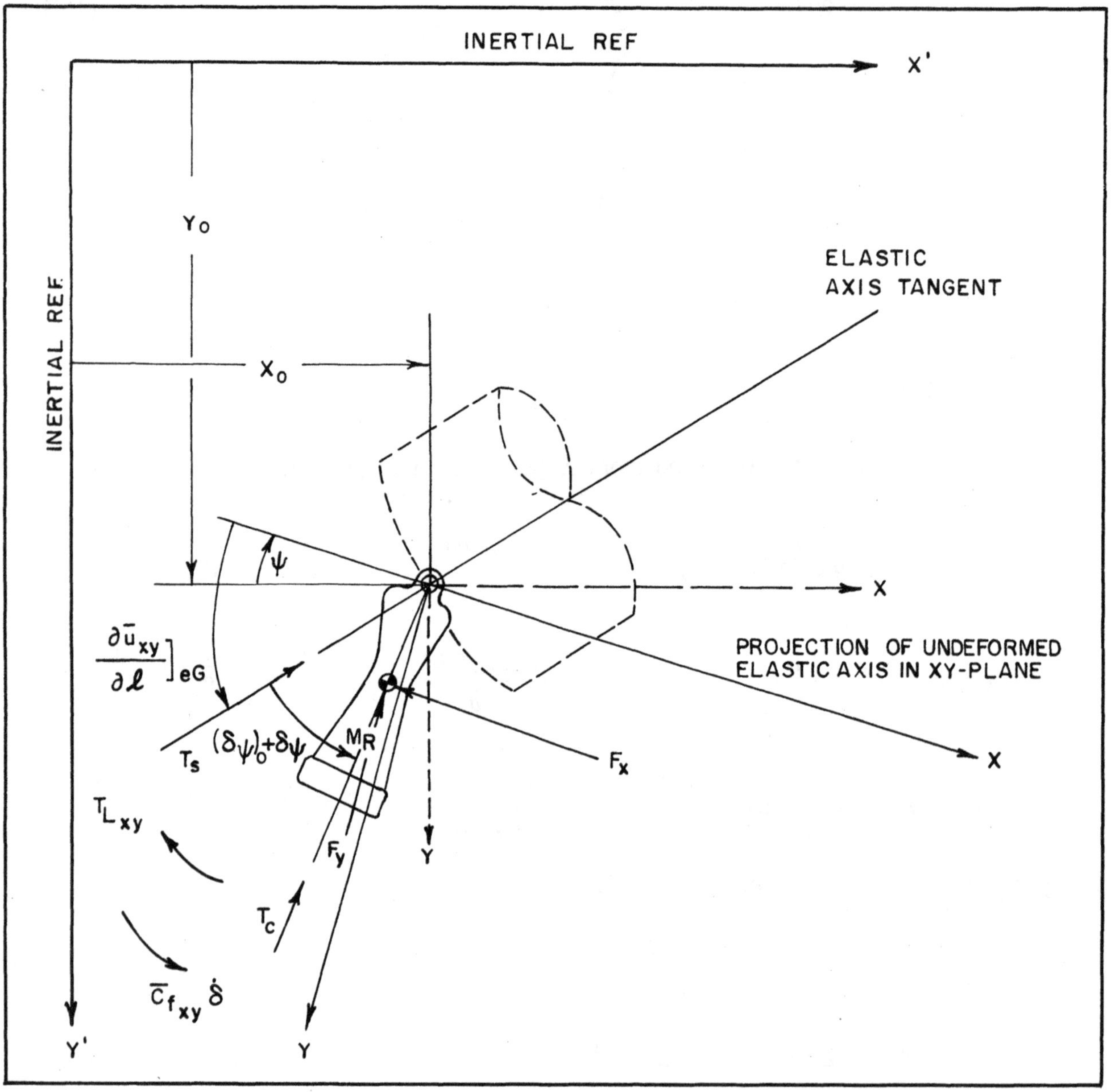

Figure 9. Control Engine in Disturbed Flight in XY-Plane

The location of the engine C.G. is given by:

$$X' = X_0 - \ell_R \cos\left((\delta_\psi)_0 + \delta_\psi + \frac{\partial \bar{u}_{xy}(eG)}{\partial \ell} - \psi \right) \tag{22B}$$

$$Y' = Y_0 + \ell_R \sin\left((\delta_\psi)_0 + \delta_\psi + \frac{\partial \bar{u}_{xy}(eG)}{\partial \ell} - \psi \right) \tag{23B}$$

The velocity components of the engine C.G. are:

$$\dot{X}' = \dot{X}_0 + \ell_R\left(\dot{\delta}_\psi + \frac{\partial^2 \bar{u}_{xy}(eG)}{\partial \ell\, \partial t} - \dot{\psi}\right)\sin\left((\delta_\psi)_0 + \delta_\psi + \frac{\partial \bar{u}_{xy}(eG)}{\partial \ell} - \psi\right)$$

$$\dot{X}' \cong \dot{X}_0$$

$$\dot{Y}' = \dot{Y}_0 + \ell_R\left(\dot{\delta}_\psi + \frac{\partial^2 \bar{u}_{xy}\,eG}{\partial \ell\, \partial t} - \dot{\psi}\right)\cos\left((\delta_\psi)_0 + \delta_\psi + \frac{\partial \bar{u}_{xy}(eG)}{\partial \ell} - \psi\right)$$

$$\dot{Y}' \cong \dot{Y}_0 + \ell_R\left(\dot{\delta}_\psi + \frac{\partial^2 \bar{u}_{xy}(eG)}{\partial \ell\, \partial t} - \dot{\psi}\right)$$

Now, $\dot{X}_0 \cong \dot{X}$, and

$$Y_0 = Y'_{CG} - \psi \ell_e + \bar{u}_{xy}(eG)\cos\psi$$

$$\dot{Y}_0 \cong \dot{Y} + \dot{\bar{u}}_{xy}(eG) - \dot{\psi}\ell_e$$

The kinetic energy is then:

$$\bar{T} = \bar{L} = \frac{M_R}{2}\left\{(\dot{X}')^2 + (\dot{Y}')^2\right\}$$

$$\bar{L} \cong \frac{M_R}{2}\left\{\dot{X}^2 + \left[\dot{Y} + \dot{\bar{u}}_{xy}(eG) - \psi\ell_e + \ell_R\left(\dot{\delta}_\psi + \frac{\partial^2 \bar{u}_{xy}(eG)}{\partial \ell\, \partial t} - \dot{\psi}\right)\right]^2\right\} \tag{24B}$$

$$\frac{\partial}{\partial t}\left(\frac{\partial \bar{L}}{\partial \dot{Y}}\right) = M_R\left[\ddot{Y} + \ddot{\bar{u}}_{xy}(eG) - \ddot{\psi}\ell_e + \ell_R\left(\ddot{\delta}_\psi + \frac{\partial^3 \bar{u}_{xy}(eG)}{\partial \ell\, \partial t^2} - \ddot{\psi}\right)\right] \tag{25B}$$

$$\frac{\partial \bar{L}}{\partial Y} = 0 \tag{26B}$$

Now, from the Lagrangean equations:

$$F_x = -M_R\ddot{X} \tag{27B}$$

$$F_y = -M_R\left[\ddot{Y} + \ddot{\bar{u}}_{xy}(eG) - \ddot{\psi}(\ell_e + \ell_R) + \ell_R\left(\ddot{\delta}_\psi + \frac{\partial^3 \bar{u}_{xy}(eG)}{\partial \ell\, \partial t^2}\right)\right] \tag{28B}$$

The moment about the missile C.G. due to these engine inertia forces is:

$$N_{ei} = F_y\left[\ell_e + \ell_R\cos\left((\delta_\psi)_0 + \delta_\psi + \frac{\partial \bar{u}_{xy}(eG)}{\partial \ell}\right)\right] \tag{29B}$$

The equation of motion of the engine about the gimbal is:

$$T_{L_{xy}} = - \left[F_x + F_y \cos \left((\delta\psi)_0 + \delta\psi + \frac{\partial \bar{u}_{xy}(eG)}{\partial \ell} \right) \right] \ell_R + \bar{C}_{f_{xy}} \dot{\delta\psi} \qquad (30B)$$

Rearranging terms and employing the modal relationships as before, Eqs. (27B), (28B), (29B), and (30B) become:

$$F_x = - M_R \left[\dot{U}_0 + \dot{u} \right] \qquad (31B)$$

$$F_y = - M_R \left\{ \dot{V}_0 + \dot{v} - (\ell_e + \ell_R) \ddot{\psi} + \ell_R \ddot{\delta}_\psi + \sum_i \left[\varphi_{xy}^{(i)}(eG) - \ell_R \, \sigma_{xy}^{(i)}(eG) \right] \ddot{q}_{xy}^{(i)} \right\} \qquad (32B)$$

$$N_{ei} = - M_R (\ell_R + \ell_e) \left\{ \dot{V}_0 + \dot{v} - (\ell_e + \ell_R) \ddot{\psi} + \ell_R \ddot{\delta}_\psi + \sum_i \left[\varphi_{xy}^{(i)}(eG) - \ell_R \, \sigma_{xy}^{(i)}(eG) \right] \ddot{q}_{xy}^{(i)} \right\} \qquad (33B)$$

$$T_{L_{xy}} = M_R \ell_R \left\{ \dot{U}_0 + \dot{u} + \dot{V}_0 + \dot{v} - (\ell_e + \ell_R) \ddot{\psi} + \ell_R \ddot{\delta}_\psi + \sum_i \left[\varphi_{xy}^{(i)}(eG) - \ell_R \, \sigma_{xy}^{(i)}(eG) \right] \ddot{q}_{xy}^{(i)} \right\} + \bar{C}_{f_{xy}} \dot{\delta}_\psi \qquad (34B)$$

Now, it can be seen from Figures 8 and 9 that the engine inertia moment acting on the missile in the rolling degree of freedom, ZY-plane, is

$$L_{ei} = F_y \, \ell_R \sin \left(\alpha_0 + (\delta\theta)_0 + \delta\theta + \frac{\partial \bar{u}_{xz}(eG)}{\partial \ell} \right) - F_z \, \ell_R \sin \left((\delta\psi)_0 + \delta\psi + \frac{\partial \bar{u}_{xy}(eG)}{\partial \ell} \right) \qquad (35B)$$

Since the expression for F_y and F_z involve only small disturbance acceleration terms, then the product of F_y or F_z times the sine of a small angle is negligible. Therefore,

$$L_{ei} \cong 0 \qquad (36B)$$

Further explanation is in order on T_L, the torque generated by the servo actuation device, and $\bar{C}_f$, the linearized, equivalent gimbal friction coefficient. $\bar{C}_f$ is a function of both viscous and coulomb friction and varies with the frequency and amplitude of engine motion. T_L is a complicated function of the servo system properties, control engine output, and autopilot command signals and is a specialized study for each control system. For the purposes of this analysis, it is only necessary to specify the general form of T_L, which will be applicable to any type of actuator. In the most general case, due to the high nonlinearity of actuators, the describing function technique used to linearize the system results in the output angle δ occurring to the third order in the expression for T_L (see Refs. 4 and 7). Thus, T_L can be written in the following general form:

$$T_{L_{xz}} = \bar{K}_1 \, \delta_{c\theta} + \bar{K}_2 \, \dddot{\delta}_\theta + \bar{K}_3 \, \ddot{\delta}_\theta + \bar{K}_4 \, \dot{\delta}_\theta + \bar{K}_5 \, \delta_\theta \qquad (37B)$$

$$T_{L_{xy}} = \bar{K}_6\, \delta_{c\psi} + \bar{K}_7\, \dddot{\delta}_\psi + \bar{K}_8\, \ddot{\delta}_\psi + \bar{K}_9\, \dot{\delta}_\psi + \bar{K}_{10}\, \delta_\psi \, , \tag{38B}$$

where the coefficients $\bar{K}_1, \cdots\cdots, \bar{K}_{10}$ are dependent on the type of servo actuation employed. Therefore, the engine equations of motion about the gimbal, Eqs. (21B) and (34B), become:

$$\begin{aligned} M_R\, \ell_R \Big\{ \dot{w} + (L - \ell_{CG})\, \ddot{\theta} + \sum_i \Big[\varphi_{xz}^{(i)}(eG) - \ell_R\, \sigma_{xz}^{(i)}(eG) \Big] \ddot{\bar{q}}_{xz}^{(i)} \\ + \dot{U}_0 \Big[\alpha_0 + (\delta_\theta)_0 - \sum_i \sigma_{xz}^{(i)}(eG)\, \bar{q}_{xz}^{(i)} \Big] \Big] \Big\} - \bar{K}_1\, \delta_{c\theta} - \bar{K}_2\, \dddot{\delta}_\theta \\ + (M_R\, \ell_R^2\, \dot{U}_0 - \bar{K}_3)\, \ddot{\delta}_\theta + (\bar{C}_{f_{xz}} - \bar{K}_4)\, \dot{\delta}_\theta + (M_R\, \ell_R\, \dot{U}_0 - \bar{K}_5)\, \delta_\theta = 0 \end{aligned} \tag{39B}$$

and

$$\begin{aligned} M_R\, \ell_R \Big\{ \dot{U}_0 + \dot{u} + \dot{V}_0 + \dot{v} - (L - \ell_{CG})\, \ddot{\psi} + \sum_i \Big[\varphi_{xy}^{(i)}(eG) - \ell_R\, \sigma_{xy}^{(i)}(eG) \Big] \ddot{\bar{q}}_{xy}^{(i)} \Big\} \\ - \bar{K}_6\, \delta_{c\psi} - \bar{K}_7\, \dddot{\delta}_\psi + (M_R\, \ell_R^2 - \bar{K}_8)\, \ddot{\delta}_\psi + (\bar{C}_{f_{xy}} - \bar{K}_9)\, \dot{\delta}_\psi - \bar{K}_{10}\, \delta_\psi = 0 \end{aligned} \tag{40B}$$

APPENDIX C

FUEL SLOSHING

APPENDIX C

FUEL SLOSHING

In this appendix the forces and moments acting on the missile due to propellant sloshing are determined; and, based on a simple spring-mass analogy, the equations of motion for the sloshing propellants are given.

It is well known that the forces produced by the sloshing propellants can be duplicated by an equivalent oscillating spring-mass system and an additional rigid mass. The location of the spring and the center of gravity of the rigid mass are dictated by the requirement for duplication of the moments produced by propellant oscillations. Only that portion of the propellants which is considered as rigid (or fixed mass) should be included in the mass distribution of the missile when the free-free elastic modes are calculated. Under this assumption, only the propellant forces and moments produced by the oscillating spring-mass must be considered as acting on the missile.

The influence of sloshing on missile dynamic stability depends, to a large extent, on the amount of internal damping which the propellants exhibit. Continuous oscillation of the control engine is the major effect of sloshing on the missile. If the propellants possess no internal natural damping, the control deflections may become large enough to couple significantly with other missile motions and cause an instability.

Figure 10 shows a schematic of a cylindrical propellant tank with the equivalent spring-mass. The oscillation modes with frequencies greater than the fundamental, or first mode, frequency possess very little energy; hence, these higher order modes and their correspondingly small spring-masses are neglected. It is assumed that sloshing occurs only in the XZ and XY-planes (no circular sloshing) and that the transverse axes of the tanks are coincident with the missile longitudinal axis.

No attempt is made here to derive the sloshing forces, moments, and equations of motion; the results to be used in the main body of this report are merely stated. The reader is referred to Refs. 7 and 10 for a complete treatment of the derivations. The expressions for the forces and moments produced by an analogous spring-mass and rigid-mass system are equivalent in form to those of the hydrodynamic equations of fluid motion. Through a term-by-term comparison of the two sets of equations, the terms in the mechanical analogy are expressed or defined in terms of the hydrodynamic parameters. As derived in Ref. 10, the equations of motion for sloshing in the ℓth tank are:

$$M_{1\ell}\ddot{Y}_{\ell_{xz}} + 2\xi_\ell\,\omega_\ell\,M_{1\ell}\dot{Y}_{\ell_{xz}} + K_\ell Y_{\ell_{xz}} = M_{1\ell}\left[(\dot{w} - \ddot{\theta}x_\ell) + \sum_i \varphi_{xz}^{(i)}(x_\ell)\,\ddot{\bar{q}}_{xz}^{(i)}\right] \tag{1C}$$

$$M_{1\ell}\ddot{Y}_{\ell_{xy}} + 2\xi_\ell\,\omega_\ell\,M_{1\ell}\dot{Y}_{\ell_{xy}} + K_\ell Y_{\ell_{xy}} = M_{1\ell}\left[(\dot{v} + \ddot{\psi}x_\ell) + \sum_i \varphi_{xy}^{(i)}(x_\ell)\,\ddot{\bar{q}}_{xy}^{(i)}\right] \tag{2C}$$

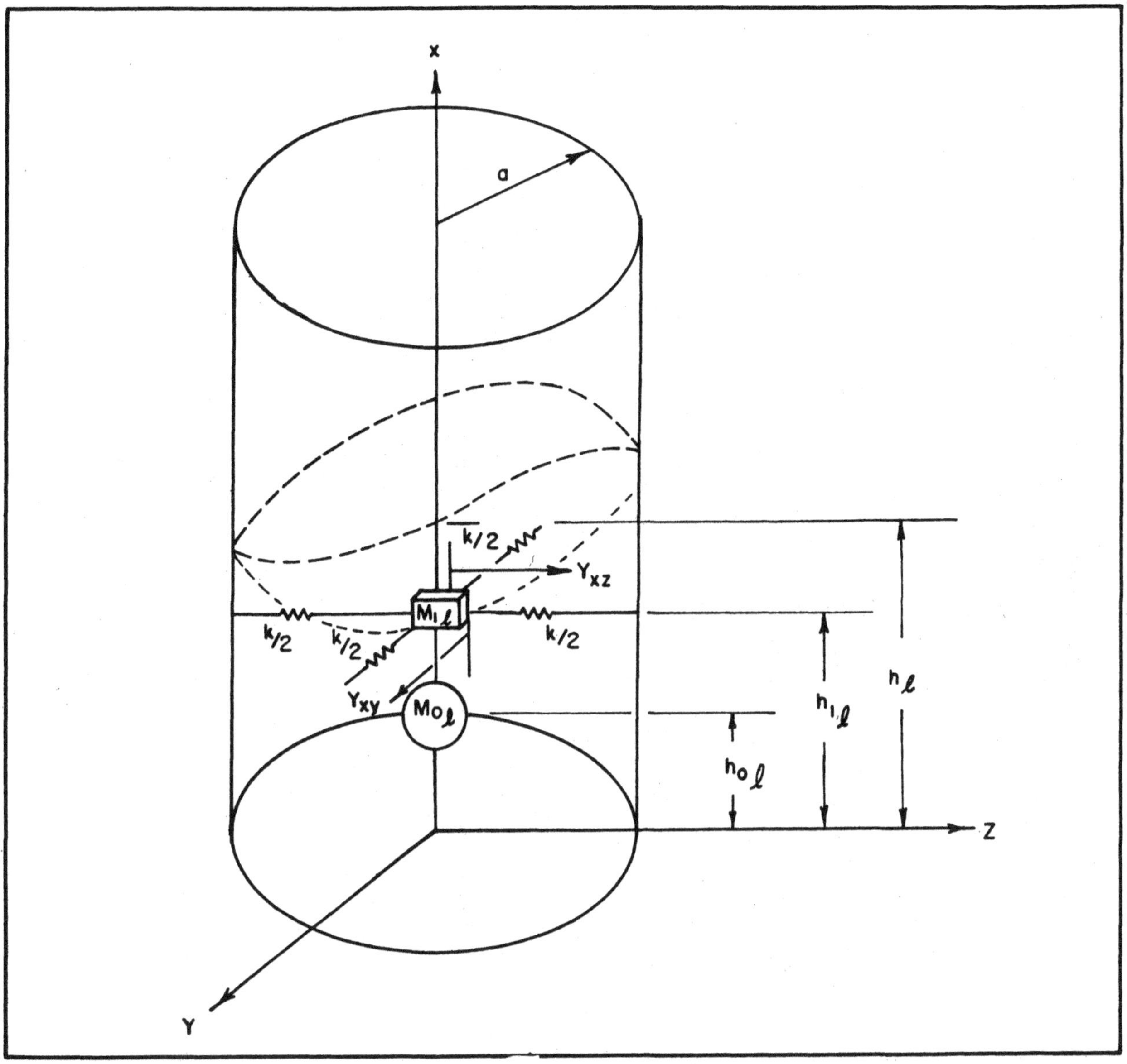

Figure 10. Sloshing Spring-Mass Analogy

The external forces and moments acting on the missile due to sloshing are:

$$F_{x_{SP}} \cong -\sum_{\ell} M_{1\ell} \, (\dot{U}_0 + \dot{u}) \tag{3C}$$

$$F_{y_{SP}} \cong -\sum_{\ell} K_{\ell} \, Y_{\ell_{xy}} \tag{4C}$$

$$F_{z_{SP}} \cong -\sum_{\ell} K_{\ell} \, Y_{\ell_{xz}} \tag{5C}$$

$$L_{SP} = 0 \quad \text{(by assumption)} \tag{6C}$$

$$M_{SP} \cong \sum_{\ell} K_{\ell} Y_{\ell_{xz}} X_{\ell} + \sum_{\ell} M_{c_{\ell}} \ddot{Y}_{\ell_{xz}} + \ddot{X} \sum_{\ell} M_{1_{\ell}} Y_{\ell_{xz}} \tag{7C}$$

$$N_{SP} \cong -\sum_{\ell} K_{\ell} Y_{\ell_{xy}} X_{\ell} - \sum_{\ell} M_{c_{\ell}} \ddot{Y}_{\ell_{xy}} - \ddot{X} \sum_{\ell} M_{1_{\ell}} Y_{\ell_{xy}} \tag{8C}$$

The term-by-term comparison yields the following expressions:

$$M_{1_{\ell}} = \frac{2 M_{P_{\ell}} \tanh K'_{\ell}}{K'_{\ell} (\bar{\xi}^2_{\ell} - 1)} \tag{9C}$$

$$\omega^2_{\ell} = \frac{8 a_{\ell} \ddot{X} \tanh K'_{\ell}}{h^2_{\ell} K'_{\ell} (\bar{\xi}^2_{\ell} - 1)} \tag{10C}$$

$$K'_{\ell} = \bar{\xi}_{\ell} \frac{h_{\ell}}{a_{\ell}} \tag{11C}$$

$\bar{\xi}_{\ell}$ is the first root of the Bessel Equation, $J'_1(\bar{\xi}_{\ell}) = 0$,

$$M_{c_{\ell}} = \frac{2 M_{P_{\ell}} h_{\ell}}{K'^2_{\ell} (\bar{\xi}^2_{\ell} - 1) \cosh K'_{\ell}} \tag{12C}$$

$$\ddot{X} \cong \frac{T_s + T_c - D}{M_t} \tag{13C}$$

$$h_{1_{\ell}} = \frac{h_{\ell} (2 + K'_{\ell} \sinh K'_{\ell} - \cosh K'_{\ell})}{K'_{\ell} \sinh K'_{\ell}} \tag{14C}$$

The portion of propellant mass in each tank considered to be rigid is then given by:

$$M_{0_{\ell}} = M_{P_{\ell}} - M_{1_{\ell}} \tag{15C}$$

And its distance from the tank bottom is:

$$h_{0_{\ell}} = \frac{\frac{1}{2} + \frac{a^2_{\ell}}{4 h^2_{\ell}} - \frac{2 (2 + K'_{\ell} \sinh K'_{\ell} - \cosh K'_{\ell}}{K'^2_{\ell} (\bar{\xi}^2_{\ell} - 1) \cosh K'_{\ell}}}{1 - \frac{2 \tanh K'_{\ell}}{K'_{\ell} (\bar{\xi}^2_{\ell} - 1)}} \tag{16C}$$

The mass moment of inertia of the rigid mass about the tank bottom is:

$$M_{0_\ell} h_{0_\ell}^2 + I_{0_\ell} = M_{P_\ell} h_\ell^2 \left[\frac{1}{3} + \frac{2(4 - K_\ell'^2) \sinh K_\ell' - 8K_\ell' + 2K_\ell' \cosh K_\ell'}{K_\ell'^3 (\bar{\xi}_\ell^2 - 1) \cosh K_\ell'} \right] \tag{17C}$$

$$K_\ell = \omega_\ell^2 M_{1_\ell} \tag{18C}$$

APPENDIX D

AERODYNAMIC FORCES

APPENDIX D

AERODYNAMIC FORCES

In this appendix the aerodynamic perturbation forces and moments are developed in terms of familiar aerodynamic coefficients and stability derivatives. Let:

$$X = \frac{1}{2} \rho \bar{U}^2 S_1 C_X \quad ,$$

where $\bar{U}$ is the relative velocity given by $\bar{U}^2 = (U_0+u)^2 + (V_0+v)^2 + w^2$.
And for $V_0 << U_0$, $\bar{U} \cong U_0$. For small α and β it can be shown that $C_X \cong C_L \alpha - C_D$. Therefore, $X \cong \frac{1}{2} \rho \bar{U}^2 S_1 \left[C_L \alpha - C_D \right]^e$.
The e indicates the coefficients are to be corrected for static aeroelastic effects.

$$X_0 \cong -\frac{1}{2} \rho U_0^2 S_1 (C_D)_0^e \tag{1D}$$

$$\left(\frac{\partial X}{\partial u}\right)_0 \cong -\frac{1}{2} \rho U_0^2 S_1 \left(\frac{\partial C_D}{\partial u}\right)_0^e - \rho U_0 S_1 (C_D)_0^e \left(\frac{\partial \bar{U}}{\partial u}\right)_0$$

Now,

$$\left(\frac{\partial \bar{U}}{\partial u}\right)_0 \cong 1 ; \quad \left(\frac{\partial \bar{U}}{\partial v}\right)_0 \cong 0 ; \quad \left(\frac{\partial \bar{U}}{\partial w}\right)_0 \cong 0 ;$$

therefore,

$$\left(\frac{\partial X}{\partial u}\right)_0 \cong -\frac{1}{2} \rho U_0^2 S_1 \left[\left(\frac{\partial C_D}{\partial u}\right)_0^e + \frac{2}{U_0} (C_D)_0^e \right] \tag{2D}$$

$$\left(\frac{\partial X}{\partial v}\right)_0 \cong -\frac{1}{2} \rho U_0^2 S_1 \left(\frac{\partial C_D}{\partial v}\right)_0^e \tag{3D}$$

$$\left(\frac{\partial X}{\partial w}\right)_0 \cong \frac{1}{2} \rho U_0^2 S_1 \left[\frac{1}{U_0} (C_L)_0^e - \left(\frac{\partial C_D}{\partial w}\right)_0^e \right] \tag{4D}$$

$$\left(\frac{\partial X}{\partial p}\right)_0 \cong -\frac{1}{2} \rho U_0^2 S_1 \left(\frac{\partial C_D}{\partial p}\right)_0^e \tag{5D}$$

$$\left(\frac{\partial X}{\partial q}\right)_0 \cong -\frac{1}{2} \rho U_0^2 S_1 \left(\frac{\partial C_D}{\partial q}\right)_0^e \tag{6D}$$

$$\left(\frac{\partial X}{\partial r}\right)_0 \cong -\frac{1}{2} \rho U_0^2 S_1 \left(\frac{\partial C_D}{\partial r}\right)_0^e \tag{7D}$$

At this point, the assumption of quasi-steady aerodynamics is introduced. This mathematical model assumes the forces to be directly proportional to and in phase with the instantaneous local angle of attack at each body station. As a result of this assumption, the acceleration derivatives of the aerodynamic forces and moments, with a few exceptions which will be mentioned as they come up, are neglected.

Now, let:

$$Y = \frac{1}{2} \rho \bar{U}^2 S_2 C_Y$$

It can also be shown that

$$C_Y \cong -C_y \cos(\beta_0 + \beta) - C_D \sin(\beta_0 + \beta),$$

or

$$C_Y \cong -C_y - C_D(\beta_0 + \beta).$$

Therefore,

$$Y \cong -\frac{1}{2} \rho \bar{U}^2 S_2 \left[C_y + C_D (\beta_0 + \beta) \right]^e$$

Then

$$Y_0 \cong -\frac{1}{2} \rho U_0^2 S_2 \left[(C_y)_0^e + (C_D)_0^e \beta_0 \right] \tag{8D}$$

$$\left(\frac{\partial Y}{\partial u}\right)_0 \cong -\frac{1}{2} \rho U_0^2 S_2 \left[\left(\frac{\partial C_y}{\partial u}\right)_0^e + \left(\frac{\partial C_D}{\partial u}\right)_0^e \beta_0 + \frac{2}{U_0} (C_y)_0^e + \frac{2}{U_0} \beta_0 (C_D)_0^e \right]. \tag{9D}$$

And since $\beta \cong \frac{v}{U_0}$,

$$\left(\frac{\partial Y}{\partial v}\right)_0 \cong -\frac{1}{2} \rho U_0^2 S_2 \left[\left(\frac{\partial C_y}{\partial v}\right)_0^e + \left(\frac{\partial C_D}{\partial v}\right)_0^e \beta_0 + \frac{1}{U_0} (C_D)_0^e \right] \tag{10D}$$

$$\left(\frac{\partial Y}{\partial w}\right)_0 \cong -\frac{1}{2} \rho U_0^2 S_2 \left[\left(\frac{\partial C_y}{\partial w}\right)_0^e + \left(\frac{\partial C_D}{\partial w}\right)_0^e \beta_0 \right] \tag{11D}$$

$$\left(\frac{\partial Y}{\partial p}\right)_0 \cong -\frac{1}{2} \rho U_0^2 S_2 \left[\left(\frac{\partial C_y}{\partial p}\right)_0^e + \left(\frac{\partial C_D}{\partial p}\right)_0^e \beta_0 \right] \tag{12D}$$

$$\left(\frac{\partial Y}{\partial q}\right)_0 \cong -\frac{1}{2} \rho U_0^2 S_2 \left[\left(\frac{\partial C_y}{\partial q}\right)_0^e + \left(\frac{\partial C_D}{\partial q}\right)_0^e \beta_0 \right] \tag{13D}$$

$$\left(\frac{\partial Y}{\partial r}\right)_0 \cong -\frac{1}{2} \rho U_0^2 S_2 \left[\left(\frac{\partial C_y}{\partial r}\right)_0^e + \left(\frac{\partial C_D}{\partial r}\right)_0^e \beta_0 \right] \tag{14D}$$

Also included in the Y derivatives is the $\left(\frac{\partial Y}{\partial \dot{v}}\right)$ derivative, which is included to account for possible sidewash effects in the XY-plane.

$$\left(\frac{\partial Y}{\partial \dot{v}}\right)_0 \cong -\frac{1}{2} \rho U^2 S_2 \left[\left(\frac{\partial C_y}{\partial \dot{v}}\right)_0^e + \left(\frac{\partial C_D}{\partial \dot{v}}\right)_0^e \beta \right] \tag{15D}$$

Similarly it can be shown that:

$$C_z \cong -(C_L + C_D \alpha) \cong -(C_L + \frac{1}{U_0} C_D w) ;$$

therefore,

$$Z \cong -\frac{1}{2} \rho \bar{U}^2 S_3 (C_L + C_D \alpha)^e .$$

Then

$$Z_0 \cong -\frac{1}{2} \rho U_0^2 S_3 (C_L)_0^e \tag{16D}$$

$$\left(\frac{\partial Z}{\partial u}\right)_0 \cong -\frac{1}{2} \rho U_0^2 S_3 \left[\left(\frac{\partial C_L}{\partial u}\right)_0^e + \frac{2}{U_0} (C_L)_0^e \right] \tag{17D}$$

$$\left(\frac{\partial Z}{\partial v}\right)_0 \cong -\frac{1}{2} \rho U_0^2 S_3 \left(\frac{\partial C_L}{\partial v}\right)_0^e \tag{18D}$$

$$\left(\frac{\partial Z}{\partial w}\right)_0 \cong -\frac{1}{2} \rho U_0^2 S_3 \left[\left(\frac{\partial C_L}{\partial w}\right)_0^e + \frac{1}{U_0} (C_D)_0^e \right] \tag{19D}$$

$$\left(\frac{\partial Z}{\partial p}\right)_0 \cong -\frac{1}{2} \rho U_0^2 S_3 \left(\frac{\partial C_L}{\partial p}\right)_0^e \tag{20D}$$

$$\left(\frac{\partial Z}{\partial q}\right)_0 \cong -\frac{1}{2} \rho U_0^2 S_3 \left(\frac{\partial C_L}{\partial q}\right)_0^e \tag{21D}$$

$$\left(\frac{\partial Z}{\partial r}\right)_0 \cong -\frac{1}{2} \rho U_0^2 S_3 \left(\frac{\partial C_L}{\partial r}\right)_0^e \tag{22D}$$

As in the case for the Y derivatives, $\left(\frac{\partial Z}{\partial \dot{w}}\right)$ is included to account for possible downwash effects in the XZ-plane:

$$\left(\frac{\partial Z}{\partial \dot{w}}\right)_0 \cong -\frac{1}{2} \rho U_0^2 S_3 \left(\frac{\partial C_L}{\partial \dot{w}}\right)_0^e \tag{23D}$$

Now, to determine the rolling moment derivatives, let

$$L = \frac{1}{2} \rho \bar{U}^2 S_4 b_1 (C_\ell)^e$$

Then:

$$L_0 \cong \frac{1}{2} \rho U_0^2 S_4 b_1 (C_\ell)_0^e \tag{24D}$$

$$\left(\frac{\partial L}{\partial u}\right)_0 \cong \frac{1}{2} \rho U_0^2 S_4 b_1 \left[\left(\frac{\partial C_\ell}{\partial u}\right)_0^e + \frac{2}{U_0} (C_\ell)_0^e\right] \tag{25D}$$

$$\left(\frac{\partial L}{\partial v}\right)_0 \cong \frac{1}{2} \rho U_0^2 S_4 b_1 \left(\frac{\partial C_\ell}{\partial v}\right)_0^e \tag{26D}$$

$$\left(\frac{\partial L}{\partial w}\right)_0 \cong \frac{1}{2} \rho U_0^2 S_4 b_1 \left(\frac{\partial C_\ell}{\partial w}\right)_0^e \tag{27D}$$

$$\left(\frac{\partial L}{\partial p}\right)_0 \cong \frac{1}{2} \rho U_0^2 S_4 b_1 \left(\frac{\partial C_\ell}{\partial p}\right)_0^e \tag{28D}$$

$$\left(\frac{\partial L}{\partial q}\right)_0 \cong \frac{1}{2} \rho U_0^2 S_4 b_1 \left(\frac{\partial C_\ell}{\partial q}\right)_0^e \tag{29D}$$

$$\left(\frac{\partial L}{\partial r}\right)_0 \cong \frac{1}{2} \rho U_0^2 S_4 b_1 \left(\frac{\partial C_\ell}{\partial r}\right)_0^e \tag{30D}$$

To determine the pitching moment derivatives, let

$$M = \frac{1}{2} \rho \overline{U}^2 S_5 b_2 (C_m)^e ;$$

therefore,

$$M_0 \cong \frac{1}{2} \rho U_0^2 S_5 b_2 (C_m)_0^e \tag{31D}$$

$$\left(\frac{\partial M}{\partial u}\right)_0 \cong \frac{1}{2} \rho U_0^2 S_5 b_2 \left[\left(\frac{\partial C_m}{\partial u}\right)_0^e + \frac{2}{U_0} (C_m)_0^e\right] \tag{32D}$$

$$\left(\frac{\partial M}{\partial v}\right)_0 \cong \frac{1}{2} \rho U_0^2 S_5 b_2 \left(\frac{\partial C_m}{\partial v}\right)_0^e \tag{33D}$$

$$\left(\frac{\partial M}{\partial w}\right)_0 \cong \frac{1}{2} \rho U_0^2 S_5 b_2 \left(\frac{\partial C_m}{\partial w}\right)_0^e \tag{34D}$$

$$\left(\frac{\partial M}{\partial p}\right)_0 \cong \frac{1}{2} \rho U_0^2 S_5 b_2 \left(\frac{\partial C_m}{\partial p}\right)_0^e \tag{35D}$$

$$\left(\frac{\partial M}{\partial q}\right)_0 \cong \frac{1}{2} \rho U_0^2 S_5 b_2 \left(\frac{\partial C_m}{\partial q}\right)_0^e \tag{36D}$$

$$\left(\frac{\partial M}{\partial r}\right)_0 \cong \frac{1}{2} \rho U_0^2 S_5 b_2 \left(\frac{\partial C_m}{\partial r}\right)_0^e \tag{37D}$$

$$\left(\frac{\partial M}{\partial \dot{w}}\right)_0 \cong \frac{1}{2} \rho U_0^2 S_5 b_2 \left(\frac{\partial C_m}{\partial \dot{w}}\right)_0^e \tag{38D}$$

To determine the yawing moment derivatives, let

$$N = \frac{1}{2} \rho \bar{U}^2 S_6 b_3 (C_n)^e ;$$

therefore,

$$N_0 \cong \frac{1}{2} \rho U_0^2 S_6 b_3 (C_n)_0^e \tag{39D}$$

$$\left(\frac{\partial N}{\partial u}\right)_0 \cong \frac{1}{2} \rho U_0^2 S_6 b_3 \left[\left(\frac{\partial C_n}{\partial u}\right)_0^e + \frac{2}{U_0} (C_n)_0^e\right] \tag{40D}$$

$$\left(\frac{\partial N}{\partial v}\right)_0 \cong \frac{1}{2} \rho U_0^2 S_6 b_3 \left(\frac{\partial C_n}{\partial v}\right)_0^e \tag{41D}$$

$$\left(\frac{\partial N}{\partial w}\right)_0 \cong \frac{1}{2} \rho U_0^2 S_6 b_3 \left(\frac{\partial C_n}{\partial w}\right)_0^e \tag{42D}$$

$$\left(\frac{\partial N}{\partial p}\right)_0 \cong \frac{1}{2} \rho U_0^2 S_6 b_3 \left(\frac{\partial C_n}{\partial p}\right)_0^e \tag{43D}$$

$$\left(\frac{\partial N}{\partial q}\right)_0 \cong \frac{1}{2} \rho U_0^2 S_6 b_3 \left(\frac{\partial C_n}{\partial q}\right)_0^e \tag{44D}$$

$$\left(\frac{\partial N}{\partial r}\right)_0 \cong \frac{1}{2} \rho U_0^2 S_6 b_3 \left(\frac{\partial C_n}{\partial r}\right)_0^e \tag{45D}$$

$$\left(\frac{\partial N}{\partial \dot{v}}\right)_0 \cong \frac{1}{2} \rho U_0^2 S_6 b_3 \left(\frac{\partial C_n}{\partial \dot{v}}\right)_0^e \tag{46D}$$

It now remains to determine the change in the aerodynamic forces and moments due to the elastic modes of the missile. The reader is referred to the elastic deflection notation introduced in Appendix A; it is used in what follows without further comment.

The local angle of attack and local side-slip angle due to elastic deflections and velocities in the respective XZ and XY-planes (see Figures 6 and 7) is given approximately by:

$$\alpha(\ell) \cong \frac{\partial \bar{u}_{xz}}{\partial \ell} + \frac{\dot{\bar{u}}_{xz}}{U_0} \tag{47D}$$

and

$$\beta(\ell) \cong \frac{\partial \bar{u}_{xy}}{\partial \ell} + \frac{\dot{\bar{u}}_{xy}}{U_0} \tag{48D}$$

Employing the notation of Appendix A, these equations can be written as

$$\alpha(\ell) \cong \sum_i \left[-\sigma_{xz}^{(i)} \bar{q}_{xz}^{(i)} + \frac{1}{U_0} \varphi_{xz}^{(i)} \dot{\bar{q}}_{xz}^{(i)} \right] \tag{49D}$$

and

$$\beta(\ell) \cong \sum_i \left[-\sigma_{xy}^{(i)} \bar{q}_{xy}^{(i)} + \frac{1}{U_0} \varphi_{xy}^{(i)} \dot{\bar{q}}_{xy}^{(i)} \right] ; \tag{50D}$$

$$\sum_i \left(\frac{\partial X}{\partial \bar{q}_{xz}^{(i)}} \right)_0 \bar{q}_{xz}^{(i)} \cong \int_0^L \left(\frac{\partial X}{\partial \alpha} \right)_0 \sum_i \left(\frac{\partial \alpha(\ell)}{\partial \bar{q}_{xz}^{(i)}} \right) \bar{q}_{xz}^{(i)} d\ell \; . \tag{51D}$$

Now, with the relation $\alpha \cong \frac{w}{U_0}$, Eq. (4D), and Eq. (49D) and based on a linear approximation, Eq. (51D) becomes:

$$\sum_i \left(\frac{\partial X}{\partial \bar{q}_{xz}^{(i)}} \right)_0 \bar{q}_{xz}^{(i)} \cong \frac{1}{2} \rho U_0^2 S_1 \int_0^L \left[(C_L(\ell))_0^e - U_0 \left(\frac{\partial C_D(\ell)}{\partial w} \right)_0^e \right] \sum_i -\sigma_{xz}^{(i)} \bar{q}_{xz}^{(i)} d\ell \tag{52D}$$

Similarly,

$$\sum_i \left(\frac{\partial X}{\partial \bar{q}_{xy}^{(i)}} \right)_0 \bar{q}_{xy}^{(i)} \cong \int_0^L \left(\frac{\partial X}{\partial \beta} \right)_0 \sum_i \left(\frac{\partial \beta}{\partial \bar{q}_{xy}^{(i)}} \right) \bar{q}_{xy}^{(i)} d\ell$$

$$\cong + \frac{1}{2} \rho U_0^3 S_1 \int_0^L \left(\frac{\partial C_D(\ell)}{\partial v} \right)_0^e \sum_i \sigma_{xy}^{(i)} \bar{q}_{xy}^{(i)} d\ell \tag{53D}$$

$$\sum_i \left(\frac{\partial X}{\partial \bar{q}_{zy}^{(i)}} \right)_0 \bar{q}_{zy}^{(i)} \cong \int_0^L \left(\frac{\partial X}{\partial p} \right)_0 \sum_i \left(\frac{\partial p}{\partial \bar{q}_{zy}^{(i)}} \right) \bar{q}_{zy}^{(i)} d\ell \cong 0 \tag{54D}$$

The remaining derivatives are derived in the same manner and are listed below.

$$\sum_i \left(\frac{\partial X}{\partial \dot{\bar{q}}_{xz}^{(i)}} \right)_0 \dot{\bar{q}}_{xz}^{(i)} \cong$$

$$\frac{1}{2} \rho U_0^2 S_1 \int_0^L \left[(C_L(\ell))_0^e - U_0 \left(\frac{\partial C_D(\ell)}{\partial w} \right)_0^e \right] \sum_i \frac{1}{U_0} \varphi_{xz}^{(i)} \dot{\bar{q}}_{xz}^{(i)} d\ell \tag{55D}$$

$$\sum_i \left(\frac{\partial X}{\partial \dot{\bar{q}}_{xy}^{(i)}} \right)_0 \dot{\bar{q}}_{xy}^{(i)} \cong -\frac{1}{2} \rho U_0^3 S_1 \int_0^L \left(\frac{\partial C_D(\ell)}{\partial v} \right)_0^e \sum_i \frac{1}{U_0} \varphi_{xy}^{(i)} \dot{\bar{q}}_{xy}^{(i)} d\ell \tag{56D}$$

$$\sum_i \left(\frac{\partial X}{\partial \dot{\bar{q}}_{zy}^{(i)}} \right)_0 \dot{\bar{q}}_{zy}^{(i)} \cong -\frac{1}{2} \rho U_0^2 S_1 \int_0^L \left(\frac{\partial C_D(\ell)}{\partial p} \right)_0^e \sum_i \varphi_{zy}^{(i)} \dot{\bar{q}}_{zy}^{(i)} d\ell \tag{57D}$$

$$\sum_i \left(\frac{\partial Y}{\partial \bar{q}_{xz}^{(i)}}\right)_0 \bar{q}_{xz}^{(i)} \cong \frac{1}{2}\rho U_0^3 S_2 \int_0^L \left[\left(\frac{\partial C_y(\ell)}{\partial w}\right)_0^e + \left(\frac{\partial C_D(\ell)}{\partial w}\right)_0^e \beta_0\right] \sum_i \sigma_{xz}^{(i)} \bar{q}_{xz}^{(i)} d\ell \tag{58D}$$

$$\sum_i \left(\frac{\partial Y}{\partial \bar{q}_{xy}^{(i)}}\right)_0 \bar{q}_{xy}^{(i)} \cong$$

$$\frac{1}{2}\rho U_0^3 S_2 \int_0^L \left[\left(\frac{\partial C_y(\ell)}{\partial v}\right)_0^e + \left(\frac{\partial C_D(\ell)}{\partial v}\right)_0^e \beta_0 + \frac{1}{U_0}\left(C_D(\ell)\right)_0^e\right] \sum_i \sigma_{xy}^{(i)} \bar{q}_{xy}^{(j)} d\ell \tag{59D}$$

$$\sum_i \left(\frac{\partial Y}{\partial \bar{q}_{zy}^{(i)}}\right)_0 \bar{q}_{zy}^{(i)} \cong 0 \tag{60D}$$

$$\sum_i \left(\frac{\partial Y}{\partial \dot{\bar{q}}_{xz}^{(i)}}\right)_0 \dot{\bar{q}}_{xz}^{(i)} \cong$$

$$-\frac{1}{2}\rho U_0^3 S_2 \int_0^L \left[\left(\frac{\partial C_y(\ell)}{\partial w}\right)_0^e + \left(\frac{\partial C_D(\ell)}{\partial w}\right)_0^e \beta_0\right] \sum_i \frac{1}{U_0} \varphi_{xz}^{(i)} \dot{\bar{q}}_{xz}^{(i)} d\ell \tag{61D}$$

$$\sum_i \left(\frac{\partial Y}{\partial \dot{\bar{q}}_{xy}^{(i)}}\right)_0 \dot{\bar{q}}_{xy}^{(i)} \cong$$

$$-\frac{1}{2}\rho U_0^3 S_2 \int_0^L \left[\left(\frac{\partial C_y(\ell)}{\partial v}\right)_0^e + \left(\frac{\partial C_D(\ell)}{\partial v}\right)_0^e \beta_0 + \frac{1}{U_0}\left(C_D(\ell)\right)_0^e\right] \sum_i \frac{1}{U_0} \varphi_{xy}^{(i)} \dot{\bar{q}}_{xy}^{(i)} d\ell \tag{62D}$$

$$\sum_i \left(\frac{\partial Y}{\partial \dot{\bar{q}}_{zy}^{(i)}}\right)_0 \dot{\bar{q}}_{zy}^{(i)} \cong -\frac{1}{2}\rho U_0^2 S_2 \int_0^L \left[\left(\frac{\partial C_y(\ell)}{\partial p}\right)_0^e + \left(\frac{\partial C_D(\ell)}{\partial p}\right)_0^e \beta_0\right] \sum_i \varphi_{zy}^{(i)} \dot{\bar{q}}_{zy}^{(i)} d\ell \tag{63D}$$

$$\sum_i \left(\frac{\partial Z}{\partial \bar{q}_{xz}^{(i)}}\right)_0 \bar{q}_{xz}^{(i)} \cong \frac{1}{2}\rho U_0^3 S_3 \int_0^L \left[\left(\frac{\partial C_L(\ell)}{\partial w}\right)_0^e + \frac{1}{U_0}\left(C_D(\ell)\right)_0^e\right] \sum_i \sigma_{xz}^{(i)} \bar{q}_{xz}^{(i)} d\ell \tag{64D}$$

$$\sum_i \left(\frac{\partial Z}{\partial \bar{q}_{xy}^{(i)}}\right)_0 \bar{q}_{xy}^{(i)} \cong \frac{1}{2}\rho U_0^3 S_3 \int_0^L \left(\frac{\partial C_L(\ell)}{\partial v}\right)_0^e \sum_i \sigma_{xy}^{(i)} \bar{q}_{xy}^{(i)} d\ell \tag{65D}$$

$$\sum_i \left(\frac{\partial Z}{\partial \bar{q}_{zy}^{(i)}}\right)_0 \bar{q}_{zy}^{(i)} \cong 0 \tag{66D}$$

$$\sum_i \left(\frac{\partial Z}{\partial \dot{\bar{q}}_{xz}^{(i)}}\right)_o \dot{\bar{q}}_{xz}^{(i)} \cong -\frac{1}{2}\rho U_o^3 S_3 \int_o^L \left[\left(\frac{\partial c_L(\ell)}{\partial w}\right)_o^e + \frac{1}{U_o}\left(c_D(\ell)\right)_o^e\right] \sum_i \frac{1}{U_o} \varphi_{xz}^{(i)} \dot{\bar{q}}_{xz}^{(i)} \, d\ell \tag{67D}$$

$$\sum_i \left(\frac{\partial Z}{\partial \dot{\bar{q}}_{xy}^{(i)}}\right)_o \dot{\bar{q}}_{xy}^{(i)} \cong -\frac{1}{2}\rho U_o^3 S_3 \int_o^L \left(\frac{\partial c_L(\ell)}{\partial v}\right)_o^e \sum_i \frac{1}{U_o} \varphi_{xy}^{(i)} \dot{\bar{q}}_{xy}^{(i)} \, d\ell \tag{68D}$$

$$\sum_i \left(\frac{\partial Z}{\partial \dot{\bar{q}}_{zy}^{(i)}}\right)_o \dot{\bar{q}}_{zy}^{(i)} \cong -\frac{1}{2}\rho U_o^2 S_3 \int_o^L \left(\frac{\partial c_L(\ell)}{\partial p}\right)_o^e \sum_i \varphi_{zy}^{(i)} \dot{\bar{q}}_{zy}^{(i)} \, d\ell \tag{69D}$$

$$\sum_i \left(\frac{\partial L}{\partial \bar{q}_{xz}^{(i)}}\right)_o \bar{q}_{xz}^{(i)} \cong -\frac{1}{2}\rho U_o^3 S_4 b_1 \int_o^L \left(\frac{\partial c_\ell(\ell)}{\partial w}\right)_o^e \sum_i \cdot \sigma_{xz}^{(i)} \bar{q}_{xz}^{(i)} \, d\ell \tag{70D}$$

$$\sum_i \left(\frac{\partial L}{\partial \bar{q}_{xy}^{(i)}}\right)_o \bar{q}_{xy}^{(i)} \cong -\frac{1}{2}\rho U_o^3 S_4 b_1 \int_o^L \left(\frac{\partial c_\ell(\ell)}{\partial v}\right)_o^e \sum_i \sigma_{xy}^{(i)} \bar{q}_{xy}^{(i)} \, d\ell \tag{71D}$$

$$\sum_i \left(\frac{\partial L}{\partial \bar{q}_{zy}^{(i)}}\right)_o \bar{q}_{zy}^{(i)} \cong 0 \tag{72D}$$

$$\sum_i \left(\frac{\partial L}{\partial \dot{\bar{q}}_{xz}^{(i)}}\right)_o \dot{\bar{q}}_{xz}^{(i)} \cong \frac{1}{2}\rho U_o^3 S_4 b_1 \int_o^L \left(\frac{\partial c_\ell(\ell)}{\partial w}\right)_o^e \sum_i \frac{1}{U_o} \varphi_{xz}^{(i)} \dot{\bar{q}}_{xz}^{(i)} \, d\ell \tag{73D}$$

$$\sum_i \left(\frac{\partial L}{\partial \dot{\bar{q}}_{xy}^{(i)}}\right)_o \dot{\bar{q}}_{xy}^{(i)} \cong \frac{1}{2}\rho U_o^3 S_4 b_1 \int_o^L \left(\frac{\partial c_\ell(\ell)}{\partial v}\right)_o^e \sum_i \frac{1}{U_o} \varphi_{xy}^{(i)} \dot{\bar{q}}_{xy}^{(i)} \, d\ell \tag{74D}$$

$$\sum_i \left(\frac{\partial L}{\partial \dot{\bar{q}}_{zy}^{(i)}}\right)_o \dot{\bar{q}}_{zy}^{(i)} \cong \frac{1}{2}\rho U_o^2 S_4 b_1 \int_o^L \left(\frac{\partial c_\ell(\ell)}{\partial p}\right)_o^e \sum_i \varphi_{zy}^{(i)} \dot{\bar{q}}_{zy}^{(i)} \, d\ell \tag{75D}$$

$$\sum_i \left(\frac{\partial M}{\partial \bar{q}_{xz}^{(i)}}\right)_o \bar{q}_{xz}^{(i)} \cong -\frac{1}{2}\rho U_o^3 S_5 b_2 \int_o^L \left(\frac{\partial c_m(\ell)}{\partial w}\right)_o^e \sum_i \sigma_{xz}^{(i)} \bar{q}_{xz}^{(i)} \, d\ell \tag{76D}$$

$$\sum_i \left(\frac{\partial M}{\partial \bar{q}_{xy}^{(i)}}\right)_o \bar{q}_{xy}^{(i)} \cong -\frac{1}{2}\rho U_o^3 S_5 b_2 \int_o^L \left(\frac{\partial c_m(\ell)}{\partial v}\right)_o^e \sum_i \sigma_{xy}^{(i)} \bar{q}_{xy}^{(i)} \, d\ell \tag{77D}$$

$$\sum_i \left(\frac{\partial M}{\partial \bar{q}_{zy}^{(i)}}\right)_o \bar{q}_{zy}^{(i)} \cong 0 \tag{78D}$$

$$\sum_i \left(\frac{\partial M}{\partial \dot{\bar{q}}_{xz}^{(i)}}\right)_0 \dot{\bar{q}}_{xz}^{(i)} \cong \frac{1}{2} \rho U_0^3 S_5 b_2 \left[\int_0^L \left(\frac{\partial C_m(\ell)}{\partial w}\right)_0^e \sum_i \frac{1}{U_0} \varphi_{xz}^{(i)} \dot{\bar{q}}_{xz}^{(i)} d\ell + \int_0^L \left(\frac{\partial C_m(\ell)}{\partial \dot{w}}\right)_0^e \sum_i - \sigma_{xz}^{(i)} \dot{\bar{q}}_{xz}^{(i)} d\ell \right] \tag{79D}$$

$$\sum_i \left(\frac{\partial M}{\partial \dot{\bar{q}}_{xy}^{(i)}}\right)_0 \dot{\bar{q}}_{xy}^{(i)} \cong \frac{1}{2} \rho U_0^3 S_5 b_2 \int_0^L \left(\frac{\partial C_m(\ell)}{\partial v}\right)_0^e \sum_i \frac{1}{U_0} \varphi_{xy}^{(i)} \dot{\bar{q}}_{xy}^{(i)} d\ell \tag{80D}$$

$$\sum_i \left(\frac{\partial M}{\partial \dot{\bar{q}}_{zy}^{(i)}}\right)_0 \dot{\bar{q}}_{zy}^{(i)} \cong \frac{1}{2} \rho U_0^2 S_5 b_2 \int_0^L \left(\frac{\partial C_m(\ell)}{\partial p}\right)_0^e \sum_i \varphi_{zy}^{(i)} \dot{\bar{q}}_{zy}^{(i)} d\ell \tag{81D}$$

$$\sum_i \left(\frac{\partial N}{\partial \bar{q}_{xz}^{(i)}}\right)_0 \bar{q}_{xz}^{(i)} \cong - \frac{1}{2} \rho U_0^3 S_6 b_3 \int_0^L \left(\frac{\partial C_n(\ell)}{\partial w}\right)_0^e \sum_i \sigma_{xz}^{(i)} \bar{q}_{xz}^{(i)} d\ell \tag{82D}$$

$$\sum_i \left(\frac{\partial N}{\partial \bar{q}_{xy}^{(i)}}\right)_0 \bar{q}_{xy}^{(i)} \cong - \frac{1}{2} \rho U_0^3 S_6 b_3 \int_0^L \left(\frac{\partial C_n(\ell)}{\partial v}\right)_0^e \sum_i \sigma_{xy}^{(i)} \bar{q}_{xy}^{(i)} d\ell \tag{83D}$$

$$\sum_i \left(\frac{\partial N}{\partial \bar{q}_{zy}^{(i)}}\right)_0 \bar{q}_{zy}^{(i)} \cong 0 \tag{84D}$$

$$\sum_i \left(\frac{\partial N}{\partial \dot{\bar{q}}_{xz}^{(i)}}\right)_0 \dot{\bar{q}}_{xz}^{(i)} \cong \frac{1}{2} \rho U_0^3 S_6 b_3 \int_0^L \left(\frac{\partial C_n(\ell)}{\partial w}\right)_0^e \sum_i \frac{1}{U_0} \varphi_{xz}^{(i)} \dot{\bar{q}}_{xz}^{(i)} d\ell \tag{85D}$$

$$\sum_i \left(\frac{\partial N}{\partial \dot{\bar{q}}_{xy}^{(i)}}\right)_0 \dot{\bar{q}}_{xy}^{(i)} \cong \frac{1}{2} \rho U_0^3 S_6 b_3 \left[\int_0^L \left(\frac{\partial C_n(\ell)}{\partial v}\right)_0^e \sum_i \frac{1}{U_0} \varphi_{xy}^{(i)} \dot{\bar{q}}_{xy}^{(i)} d\ell - \int_0^L \left(\frac{\partial C_n(\ell)}{\partial \dot{v}}\right)_0^e \sum_i \sigma_{xy}^{(i)} \dot{\bar{q}}_{xy}^{(i)} d\ell \right] \tag{86D}$$

$$\sum_i \left(\frac{\partial N}{\partial \dot{\bar{q}}_{zy}^{(i)}}\right)_0 \dot{\bar{q}}_{zy}^{(i)} \cong \frac{1}{2} \rho U_0^2 S_6 b_3 \int_0^L \left(\frac{\partial C_n(\ell)}{\partial p}\right)_0^e \sum_i \varphi_{zy}^{(i)} \dot{\bar{q}}_{zy}^{(i)} d\ell \tag{87D}$$

APPENDIX E

DETERMINANT ELEMENTS

APPENDIX E

DETERMINANT ELEMENTS

$$A_{1,1} = \left\{ 1 + \frac{M_R}{\mathcal{M}_{xz}^{(1)}} \left(\varphi_{xz}^{(1)}(eG) - \ell_R \, \sigma_{xz}^{(1)}(eG) \right) \varphi_{xz}^{(1)}(\ell_R) \right\} s^2 + \left\{ 2 \xi_{xz}^{(1)} \omega_{xz}^{(1)} \right.$$

$$\left. + \frac{\rho U_0^2 S_3}{2 \mathcal{M}_{xz}^{(1)}} \int_0^L \left[\left(\frac{\partial C_L(\ell)}{\partial w} \right)_0^e + \frac{1}{U_0} C_D(\ell)_0^e \right] \varphi_{xz}^{(1)2} d\ell \right\} s + \left\{ \omega_{xz}^{(1)2} \right.$$

$$\left. - \frac{(T_s + T_c)}{\mathcal{M}_{xz}^{(1)}} \sigma_{xz}^{(1)}(eG) \, \varphi_{xz}^{(1)}(eG) - \frac{\rho U_0^3 S_3}{2 \mathcal{M}_{xz}^{(1)}} \int_0^L \left[\left(\frac{\partial C_L(\ell)}{\partial w} \right)_0^e + \frac{1}{U_0} C_D(\ell)_0^e \right] \sigma_{xz}^{(1)} \varphi_{xz}^{(1)} d\ell \right\} .$$

$$A_{1,2} = \left\{ \frac{M_R}{\mathcal{M}_{xz}^{(1)}} \left(\varphi_{xz}^{(2)}(eG) - \ell_R \, \sigma_{xz}^{(2)}(eG) \right) \varphi_{xz}^{(1)}(\ell_R) \right\} s^2 + \left\{ \frac{\rho U_0^2 S_3}{2 \mathcal{M}_{xz}^{(1)}} \int_0^L \left[\left(\frac{\partial C_L(\ell)}{\partial w} \right)_0^e \right. \right.$$

$$\left. \left. + \frac{1}{U_0} C_D(\ell)_0^e \right] \varphi_{xz}^{(2)} \varphi_{xz}^{(1)} d\ell \right\} s + \left\{ - \frac{(T_s + T_c)}{\mathcal{M}_{xz}^{(1)}} \sigma_{xz}^{(2)}(eG) \, \varphi_{xz}^{(1)}(eG) \right.$$

$$\left. - \frac{\rho U_0^3 S_3}{2 \mathcal{M}_{xz}^{(1)}} \int_0^L \left[\left(\frac{\partial C_L(\ell)}{\partial w} \right)_0^e + \frac{1}{U_0} C_D(\ell)_0^e \right] \sigma_{xz}^{(2)} \varphi_{xz}^{(1)} d\ell \right\} .$$

$$A_{1,3} = \left\{ \frac{M_R}{\mathcal{M}_{xz}^{(1)}} \left(\varphi_{xz}^{(3)}(eG) - \ell_R \, \sigma_{xz}^{(3)}(eG) \right) \varphi_{xz}^{(1)}(\ell_R) \right\} s^2 + \left\{ \frac{\rho U_0^2 S_3}{2 \mathcal{M}_{xz}^{(1)}} \int_0^L \left[\left(\frac{\partial C_L(\ell)}{\partial w} \right)_0^e \right. \right.$$

$$\left. \left. + \frac{1}{U_0} C_D(\ell)_0^e \right] \varphi_{xz}^{(3)} \varphi_{xz}^{(1)} d\ell \right\} s + \left\{ - \frac{(T_s + T_c)}{\mathcal{M}_{xz}^{(1)}} \sigma_{xz}^{(3)}(eG) \, \varphi_{xz}^{(1)}(eG) \right.$$

$$\left. - \frac{\rho U_0^3 S_3}{2 \mathcal{M}_{xz}^{(1)}} \int_0^L \left[\left(\frac{\partial C_L(\ell)}{\partial w} \right)_0^e + \frac{1}{U_0} C_D(\ell)_0^e \right] \sigma_{xz}^{(3)} \varphi_{xz}^{(1)} d\ell \right\} .$$

$$A_{1,4} = \left\{ \frac{M_R}{\mathcal{M}_{xz}^{(1)}} \left(\varphi_{xz}^{(4)}(eG) - \ell_R \, \sigma_{xz}^{(4)}(eG) \right) \varphi_{xz}^{(1)}(\ell_R) \right\} s^2 + \left\{ \frac{\rho U_0^2 S_3}{2\mathcal{M}_{xz}^{(1)}} \int_0^L \left[\left(\frac{\partial c_L(\ell)}{\partial w} \right)_0^e \right.\right.$$

$$\left.\left. + \frac{1}{U_0} \, c_D(\ell)_0^e \right] \varphi_{xz}^{(4)} \varphi_{xz}^{(1)} d\ell \right\} s + \left\{ - \frac{(T_s + T_c)}{\mathcal{M}_{xz}^{(1)}} \, \sigma_{xz}^{(4)}(eG) \, \varphi_{xz}^{(1)}(eG) \right.$$

$$\left. - \frac{\rho U_0^3 S_3}{2\mathcal{M}_{xz}^{(1)}} \int_0^L \left[\left(\frac{\partial c_L(\ell)}{\partial w} \right)_0^e + \frac{1}{U_0} \, c_D(\ell)_0^e \right] \sigma_{xz}^{(4)} \, \varphi_{xz}^{(1)} \, d\ell \right\} .$$

$$A_{1,5} = \left\{ \frac{M_R}{\mathcal{M}_{xz}^{(1)}} \left(\varphi_{xz}^{(5)}(eG) - \ell_R \, \sigma_{xz}^{(5)}(eG) \right) \varphi_{xz}^{(1)}(\ell_R) \right\} s^2 + \left\{ \frac{\rho U_0^2 S_3}{2\mathcal{M}_{xz}^{(1)}} \int_0^L \left[\left(\frac{\partial c_L(\ell)}{\partial w} \right)_0^e \right.\right.$$

$$\left.\left. + \frac{1}{U_0} \, c_D(\ell)_0^e \right] \varphi_{xz}^{(5)} \varphi_{xz}^{(1)} d\ell \right\} s + \left\{ - \frac{(T_s + T_c)}{\mathcal{M}_{xz}^{(1)}} \, \sigma_{xz}^{(5)}(eG) \, \varphi_{xz}^{(1)}(eG) \right.$$

$$\left. - \frac{\rho U_0^3 S_3}{2\mathcal{M}_{xz}^{(1)}} \int_0^L \left[\left(\frac{\partial c_L(\ell)}{\partial w} \right)_0^e + \frac{1}{U_0} \, c_D(\ell)_0^e \right] \sigma_{xz}^{(5)} \, \varphi_{xz}^{(1)} \, d\ell \right\} .$$

$$A_{1,6} = \left\{ \frac{\rho U_0^2 S_3}{2\mathcal{M}_{xz}^{(1)}} \int_0^L \left(\frac{\partial c_L(\ell)}{\partial v} \right)_0^e \varphi_{xy}^{(1)} \, \varphi_{xz}^{(1)} \, d\ell \right\} s$$

$$+ \left\{ - \frac{\rho U_0^3 S_3}{2\mathcal{M}_{xz}^{(1)}} \int_0^L \left(\frac{\partial c_L(\ell)}{\partial v} \right)_0^e \sigma_{xy}^{(1)} \, \varphi_{xz}^{(1)} \, d\ell \right\} .$$

$$A_{1,7} = \left\{ \frac{\rho U_0^2 S_3}{2\mathcal{M}_{xz}^{(1)}} \int_0^L \left(\frac{\partial c_L(\ell)}{\partial v} \right)_0^e \varphi_{xy}^{(2)} \, \varphi_{xz}^{(1)} \, d\ell \right\} s$$

$$+ \left\{ - \frac{\rho U_0^3 S_3}{2\mathcal{M}_{xz}^{(1)}} \int_0^L \left(\frac{\partial c_L(\ell)}{\partial v} \right)_0^e \sigma_{xy}^{(2)} \, \varphi_{xz}^{(1)} \, d\ell \right\} .$$

$$A_{1,8} = \left\{ \frac{\rho U_0^2 S_3}{2\mathcal{M}_{xz}^{(1)}} \int_0^L \left(\frac{\partial c_L(\ell)}{\partial v} \right)_0^e \varphi_{xy}^{(3)} \, \varphi_{xz}^{(1)} \, d\ell \right\} s$$

$$+ \left\{ - \frac{\rho U_0^3 S_3}{2\mathcal{M}_{xz}^{(1)}} \int_0^L \left(\frac{\partial c_L(\ell)}{\partial v} \right)_0^e \sigma_{xy}^{(3)} \, \varphi_{xz}^{(1)} \, d\ell \right\} .$$

$$A_{1,9} = \left\{ \frac{\rho U_0^2 S_3}{2 \mathcal{M}_{xz}^{(1)}} \int_0^L \left(\frac{\partial C_L(\ell)}{\partial v} \right)_0^e \varphi_{xy}^{(4)} \varphi_{xz}^{(1)} d\ell \right\} s$$

$$+ \left\{ - \frac{\rho U_0^3 S_3}{2 \mathcal{M}_{xz}^{(1)}} \int_0^L \left(\frac{\partial C_L(\ell)}{\partial v} \right)_0^e \sigma_{xy}^{(4)} \varphi_{xz}^{(1)} d\ell \right\} .$$

$$A_{1,10} = \left\{ \frac{\rho U_0^2 S_3}{2 \mathcal{M}_{xz}^{(1)}} \int_0^L \left(\frac{\partial C_L(\ell)}{\partial v} \right)_0^e \varphi_{xy}^{(5)} \varphi_{xz}^{(1)} d\ell \right\} s$$

$$+ \left\{ - \frac{\rho U_0^3 S_3}{2 \mathcal{M}_{xz}^{(1)}} \int_0^L \left(\frac{\partial C_L(\ell)}{\partial v} \right)_0^e \sigma_{xy}^{(5)} \varphi_{xz}^{(1)} d\ell \right\} .$$

$$A_{1,11} = \left\{ \frac{\rho U_0^2 S_3}{2 \mathcal{M}_{xz}^{(1)}} \int_0^L \left(\frac{\partial C_L(\ell)}{\partial p} \right)_0^e \varphi_{zy}^{(1)} \varphi_{xz}^{(1)} d\ell \right\} s .$$

$$A_{1,12} = \left\{ \frac{\rho U_0^2 S_3}{2 \mathcal{M}_{xz}^{(1)}} \int_0^L \left(\frac{\partial C_L(\ell)}{\partial p} \right)_0^e \varphi_{zy}^{(2)} \varphi_{xz}^{(1)} d\ell \right\} s$$

$$A_{1,13} = \left\{ \frac{\rho U_0^2 S_3}{2 \mathcal{M}_{xz}^{(1)}} \int_0^L \left(\frac{\partial C_L(\ell)}{\partial p} \right)_0^e \varphi_{zy}^{(3)} \varphi_{xz}^{(1)} d\ell \right\} s$$

$$A_{1,14} = \left\{ \frac{\rho U_0^2 S_3}{2 \mathcal{M}_{xz}^{(1)}} \int_0^L \left(\frac{\partial C_L(\ell)}{\partial p} \right)_0^e \varphi_{zy}^{(4)} \varphi_{xz}^{(1)} d\ell \right\} s$$

$$A_{1,15} = \left\{ \frac{\rho U_0^2 S_3}{2 \mathcal{M}_{xz}^{(1)}} \int_0^L \left(\frac{\partial C_L(\ell)}{\partial p} \right)_0^e \varphi_{zy}^{(5)} \varphi_{xz}^{(1)} d\ell \right\} s$$

$$A_{1,16} = \left\{ \frac{\rho U_0^2 S_3}{2 \mathcal{M}_{xz}^{(1)}} \int_0^L \left[\left(\frac{\partial C_L(\ell)}{\partial u} \right)_0^e + \frac{2}{U_0} C_L(\ell)_0^e \right] \varphi_{xz}^{(1)} d\ell \right\} .$$

$$A_{1,17} = \left\{ \frac{\rho U_0^2 S_3}{2 \mathcal{M}_{xz}^{(1)}} \int_0^L \left(\frac{\partial C_L(\ell)}{\partial v} \right)_0^e \varphi_{xz}^{(1)} d\ell \right\} .$$

$$A_{1'18} = \left\{ \frac{\rho U_0^2 S_3}{2\mathcal{M}_{xz}^{(1)}} \int_0^L \left(\frac{\partial C_L(\ell)}{\partial \dot{w}} \right)_0^e \varphi_{xz}^{(1)} d\ell + \frac{M_R}{\mathcal{M}_{xz}^{(1)}} \varphi_{xz}^{(1)}(\ell_R) \right\} s$$

$$+ \left\{ \frac{\rho U_0^2 S_3}{2\mathcal{M}_{xz}^{(1)}} \int_0^L \left[\left(\frac{\partial C_L(\ell)}{\partial w} \right)_0^e + \frac{1}{U_0} C_D(\ell)_0^e \right] \varphi_{xz}^{(1)} d\ell \right\}.$$

$$A_{1'19} = \left\{ \frac{\rho U_0^2 S_3}{2\mathcal{M}_{xz}^{(1)}} \int_0^L \left(\frac{\partial C_L(\ell)}{\partial p} \right)_0^e \varphi_{xz}^{(1)} d\ell \right\} s$$

$$A_{1'20} = \left\{ \frac{M_R}{\mathcal{M}_{xz}^{(1)}} (L - \ell_{CG}) \varphi_{xz}^{(1)}(\ell_R) \right\} s^2 + \left\{ \frac{\rho U_0^2 S_3}{2\mathcal{M}_{xz}^{(1)}} \int_0^L \left(\frac{\partial C_L(\ell)}{\partial q} \right)_0^e \varphi_{xz}^{(1)} d\ell \right\} s$$

$$A_{1'21} = \left\{ \frac{\rho U_0^2 S_3}{2\mathcal{M}_{xz}^{(1)}} \int_0^L \left(\frac{\partial C_L(\ell)}{\partial r} \right)_0^e \varphi_{xz}^{(1)} d\ell \right\} s.$$

$$A_{1'22} = \left\{ \frac{-M_{c_1}}{\mathcal{M}_{xz}^{(1)}} \sigma_{xz}^{(1)}(X_{1B}) \right\} s^2 + \left\{ \frac{K_1}{\mathcal{M}_{xz}^{(1)}} \varphi_{xz}^{(1)}(X_1) - \frac{M_{1_1}(T_s + T_c - D)}{M_t \mathcal{M}_{xz}^{(1)}} \sigma_{xz}^{(1)}(X_{1B}) \right\}$$

$$A_{1'23} = \left\{ \frac{-M_{c_2}}{\mathcal{M}_{xz}^{(1)}} \sigma_{xz}^{(1)}(X_{2B}) \right\} s^2 + \left\{ \frac{K_2}{\mathcal{M}_{xz}^{(1)}} \varphi_{xz}^{(1)}(X_2) - \frac{M_{1_2}(T_s + T_c - D)}{M_t \mathcal{M}_{xz}^{(1)}} \sigma_{xz}^{(1)}(X_{2B}) \right\}$$

$$A_{1'24} = \left\{ \frac{-M_{c_3}}{\mathcal{M}_{xz}^{(1)}} \sigma_{xz}^{(1)}(X_{3B}) \right\} s^2 + \left\{ \frac{K_3}{\mathcal{M}_{xz}^{(1)}} \varphi_{xz}^{(1)}(X_3) - \frac{M_{1_3}(T_s + T_c - D)}{M_t \mathcal{M}_{xz}^{(1)}} \sigma_{xz}^{(1)}(X_{3B}) \right\}$$

$$A_{1'25} = \left\{ \frac{-M_{c_4}}{\mathcal{M}_{xz}^{(1)}} \sigma_{xz}^{(1)}(X_{4B}) \right\} s^2 + \left\{ \frac{K_4}{\mathcal{M}_{xz}^{(1)}} \varphi_{xz}^{(1)}(X_4) - \frac{M_{1_4}(T_s + T_c - D)}{M_t \mathcal{M}_{xz}^{(1)}} \sigma_{xz}^{(1)}(X_{4B}) \right\}$$

$$A_{1'26} = A_{1'27} = A_{1'28} = A_{1'29} = 0.$$

$$A_{1,30} = \left\{ \frac{M_R}{\mathcal{M}_{xz}^{(1)}} \ell_R \varphi_{xz}^{(1)}(\ell_R) \right\} s^2 + \left\{ \frac{T_c}{\mathcal{M}_{xz}^{(1)}} \varphi_{xz}^{(1)}(eG) \right\}.$$

$$A_{1,31} = A_{1,32} = A_{1,33} = A_{1,34} = A_{1,35} = A_{1,36} = A_{1,37} = A_{1,38} = A_{1,39} = A_{1,40} = 0$$

$$A_{2,1} = \left\{ \frac{M_R}{\mathcal{M}_{xz}^{(2)}} \left(\varphi_{xz}^{(1)}(eG) - \ell_R \sigma_{xz}^{(1)}(eG)\right) \varphi_{xz}^{(2)}(\ell_R) \right\} s^2 + \left\{ \frac{\rho U_0^2 S_3}{2\mathcal{M}_{xz}^{(2)}} \int_0^L \left[\left(\frac{\partial C_L(\ell)}{\partial w}\right)_0^e \right.\right.$$

$$\left.\left. + \frac{1}{U_0} C_D(\ell)_0^e \right] \varphi_{xz}^{(1)} \varphi_{xz}^{(2)} d\ell \right\} s + \left\{ - \frac{(T_s + T_c)}{\mathcal{M}_{xz}^{(2)}} \sigma_{xz}^{(1)}(eG) \varphi_{xz}^{(2)}(eG) \right.$$

$$\left. - \frac{\rho U_0^3 S_3}{2\mathcal{M}_{xz}^{(2)}} \int_0^L \left[\left(\frac{\partial C_L(\ell)}{\partial w}\right)_0^e + \frac{1}{U_0} C_D(\ell)_0^e \right] \sigma_{xz}^{(1)} \varphi_{xz}^{(2)} d\ell \right\}.$$

$$A_{2,2} = \left\{ 1 + \frac{M_R}{\mathcal{M}_{xz}^{(2)}} \left(\varphi_{xz}^{(2)}(eG) - \ell_R \sigma_{xz}^{(2)}(eG)\right) \varphi_{xz}^{(2)}(\ell_R) \right\} s^2 \left\{ 2\xi_{xz}^{(2)} \omega_{xz}^{(2)} \right.$$

$$\left. + \frac{\rho U_0^2 S_3}{2\mathcal{M}_{xz}^{(2)}} \int_0^L \left[\left(\frac{\partial C_L(\ell)}{\partial w}\right)_0^e + \frac{1}{U_0} C_D(\ell)_0^e \right] \varphi_{xz}^{(2)2} d\ell \right\} s + \left\{ \omega_{xz}^{(2)2} \right.$$

$$\left. - \frac{(T_s + T_c)}{\mathcal{M}_{xz}^{(2)}} \sigma_{xz}^{(2)}(eG) \varphi_{xz}^{(2}(eG) - \frac{\rho U_0^3 S_3}{2\mathcal{M}_{xz}^{(2)}} \int_0^L \left[\left(\frac{\partial C_L(\ell)}{\partial w}\right)_0^e + \frac{1}{U_0} C_D(\ell)_0^e \right] \sigma_{xz}^{(2)} \varphi_{xz}^{(2)} d\ell \right\}.$$

$$A_{2,3} = \left\{ \frac{M_R}{\mathcal{M}_{xz}^{(2)}} \left(\varphi_{xz}^{(3)}(eG) - \ell_R \sigma_{xz}^{(3)}(eG)\right) \varphi_{xz}^{(2)}(\ell_R) \right\} s^2 + \left\{ \frac{\rho U_0^2 S_3}{2\mathcal{M}_{xz}^{(2)}} \int_0^L \left[\left(\frac{\partial C_L(\ell)}{\partial w}\right)_0^e \right.\right.$$

$$\left.\left. + \frac{1}{U_0} C_D(\ell)_0^e \right] \varphi_{xz}^{(3)} \varphi_{xz}^{(2)} d\ell \right\} s + \left\{ - \frac{(T_s + T_c)}{\mathcal{M}_{xz}^{(2)}} \sigma_{xz}^{(3)}(eG) \varphi_{xz}^{(2)}(eG) \right.$$

$$\left. - \frac{\rho U_0^3 S_3}{2\mathcal{M}_{xz}^{(2)}} \int_0^L \left[\left(\frac{\partial C_L(\ell)}{\partial w}\right)_0^e + \frac{1}{U_0} C_D(\ell)_0^e \right] \sigma_{xz}^{(3)} \varphi_{xz}^{(2)} d\ell \right\}.$$

$$A_{2,4} = \left\{ \frac{M_R}{\mathcal{M}_{xz}^{(2)}} \left(\varphi_{xz}^{(4)}(eG) - \ell_R \, \sigma_{xz}^{(4)}(eG) \right) \varphi_{xz}^{(2)}(\ell_R) \right\} s^2 + \left\{ \frac{\rho U_0^2 S_3}{2\mathcal{M}_{xz}^{(2)}} \int_0^L \left[\left(\frac{\partial C_L(\ell)}{\partial w} \right)_0^e \right.\right.$$

$$\left.\left. + \frac{1}{U_0} C_D(\ell)_0^e \right] \varphi_{xz}^{(4)} \varphi_{xz}^{(2)} d\ell \right\} s + \left\{ - \frac{(T_s + T_c)}{\mathcal{M}_{xz}^{(2)}} \sigma_{xz}^{(4)}(eG) \, \varphi_{xz}^{(2)}(eG) \right.$$

$$\left. - \frac{\rho U_0^3 S_3}{2\mathcal{M}_{xz}^{(2)}} \int_0^L \left[\left(\frac{\partial C_L(\ell)}{\partial w} \right)_0^e + \frac{1}{U_0} C_D(\ell)_0^e \right] \sigma_{xz}^{(4)} \varphi_{xz}^{(2)} d\ell \right\} .$$

$$A_{2,5} = \left\{ \frac{M_R}{\mathcal{M}_{xz}^{(2)}} \left(\varphi_{xz}^{(5)}(eG) - \ell_R \, \sigma_{xz}^{(5)}(eG) \right) \varphi_{xz}^{(2)}(\ell_R) \right\} s^2 + \left\{ \frac{\rho U_0^2 S_3}{2\mathcal{M}_{xz}^{(2)}} \int_0^L \left[\left(\frac{\partial C_L(\ell)}{\partial w} \right)_0^e \right.\right.$$

$$\left.\left. + \frac{1}{U_0} C_D(\ell)_0^e \right] \varphi_{xz}^{(5)} \varphi_{xz}^{(2)} d\ell \right\} s + \left\{ - \frac{(T_s + T_c)}{\mathcal{M}_{xz}^{(2)}} \sigma_{xz}^{(5)}(eG) \, \varphi_{xz}^{(2)}(eG) \right.$$

$$\left. - \frac{\rho U_0^3 S_3}{2\mathcal{M}_{xz}^{(2)}} \int_0^L \left[\left(\frac{\partial C_L(\ell)}{\partial w} \right)_0^e + \frac{1}{U_0} C_D(\ell)_0^e \right] \sigma_{xz}^{(5)} \varphi_{xz}^{(2)} d\ell \right\} .$$

$$A_{2,6} = \left\{ \frac{\rho U_0^2 S_3}{2\mathcal{M}_{xz}^{(2)}} \int_0^L \left(\frac{\partial C_L(\ell)}{\partial v} \right)_0^e \varphi_{xy}^{(1)} \varphi_{xz}^{(2)} d\ell \right\} s$$

$$+ \left\{ - \frac{\rho U_0^3 S_3}{2\mathcal{M}_{xz}^{(2)}} \int_0^L \left(\frac{\partial C_L(\ell)}{\partial v} \right)_0^e \sigma_{xy}^{(1)} \varphi_{xz}^{(2)} d\ell \right\} .$$

$$A_{2,7} = \left\{ \frac{\rho U_0^2 S_3}{2\mathcal{M}_{xz}^{(2)}} \int_0^L \left(\frac{\partial C_L(\ell)}{\partial v} \right)_0^e \varphi_{xy}^{(2)} \varphi_{xz}^{(2)} d\ell \right\} s$$

$$+ \left\{ - \frac{\rho U_0^3 S_3}{2\mathcal{M}_{xz}^{(2)}} \int_0^L \left(\frac{\partial C_L(\ell)}{\partial v} \right)_0^e \sigma_{xy}^{(2)} \varphi_{xz}^{(2)} d\ell \right\} .$$

$$A_{2,8} = \left\{ \frac{\rho U_0^2 S_3}{2\mathcal{M}_{xz}^{(2)}} \int_0^L \left(\frac{\partial C_L(\ell)}{\partial v} \right)_0^e \varphi_{xy}^{(3)} \varphi_{xz}^{(2)} d\ell \right\} s$$

$$+ \left\{ - \frac{\rho U_0^3 S_3}{2\mathcal{M}_{xz}^{(2)}} \int_0^L \left(\frac{\partial C_L(\ell)}{\partial v} \right)_0^e \sigma_{xy}^{(3)} \varphi_{xz}^{(2)} d\ell \right\} .$$

$$A_{2,9} = \left\{ \frac{\rho U_0^2 S_3}{2\mathcal{M}_{xz}^{(2)}} \int_0^L \left(\frac{\partial C_L(\ell)}{\partial v}\right)_0^e \varphi_{xy}^{(4)} \varphi_{xz}^{(2)} d\ell \right\} s$$

$$+ \left\{ - \frac{\rho U_0^3 S_3}{2\mathcal{M}_{xz}^{(2)}} \int_0^L \left(\frac{\partial C_L(\ell)}{\partial v}\right)_0^e \sigma_{xy}^{(4)} \varphi_{xz}^{(2)} d\ell \right\}.$$

$$A_{2,10} = \left\{ \frac{\rho U_0^2 S_3}{2\mathcal{M}_{xz}^{(2)}} \int_0^L \left(\frac{\partial C_L(\ell)}{\partial v}\right)_0^e \varphi_{xy}^{(5)} \varphi_{xz}^{(2)} d\ell \right\} s$$

$$+ \left\{ - \frac{\rho U_0^3 S_3}{2\mathcal{M}_{xz}^{(2)}} \int_0^L \left(\frac{\partial C_L(\ell)}{\partial v}\right)_0^e \sigma_{xy}^{(5)} \varphi_{xz}^{(2)} d\ell \right\}.$$

$$A_{2,11} = \left\{ \frac{\rho U_0^2 S_3}{2\mathcal{M}_{xz}^{(2)}} \int_0^L \left(\frac{\partial C_L(\ell)}{\partial p}\right)_0^e \varphi_{zy}^{(1)} \varphi_{xz}^{(2)} d\ell \right\} s.$$

$$A_{2,12} = \left\{ \frac{\rho U_0^2 S_3}{2\mathcal{M}_{xz}^{(2)}} \int_0^L \left(\frac{\partial C_L(\ell)}{\partial p}\right)_0^e \varphi_{zy}^{(2)} \varphi_{xz}^{(2)} d\ell \right\} s.$$

$$A_{2,13} = \left\{ \frac{\rho U_0^2 S_3}{2\mathcal{M}_{xz}^{(2)}} \int_0^L \left(\frac{\partial C_L(\ell)}{\partial p}\right)_0^e \varphi_{zy}^{(3)} \varphi_{xz}^{(2)} d\ell \right\} s.$$

$$A_{2,14} = \left\{ \frac{\rho U_0^2 S_3}{2\mathcal{M}_{xz}^{(2)}} \int_0^L \left(\frac{\partial C_L(\ell)}{\partial p}\right)_0^e \varphi_{zy}^{(4)} \varphi_{xz}^{(2)} d\ell \right\} s.$$

$$A_{2,15} = \left\{ \frac{\rho U_0^2 S_3}{2\mathcal{M}_{xz}^{(2)}} \int_0^L \left(\frac{\partial C_L(\ell)}{\partial p}\right)_0^e \varphi_{zy}^{(5)} \varphi_{xz}^{(2)} d\ell \right\} s.$$

$$A_{2,16} = \left\{ \frac{\rho U_0^2 S_3}{2\mathcal{M}_{xz}^{(2)}} \int_0^L \left[\left(\frac{\partial C_L(\ell)}{\partial u}\right)_0^e + \frac{2}{U_0} C_L(\ell)_0^e \right] \varphi_{xz}^{(2)} d\ell \right\}.$$

$$A_{2,17} = \left\{ \frac{\rho U_0^2 S_3}{2\mathcal{M}_{xz}^{(2)}} \int_0^L \left(\frac{\partial C_L(\ell)}{\partial v}\right)_0^e \varphi_{xz}^{(2)} d\ell \right\}.$$

$$A_{2,18} = \left\{ \frac{\rho U_0^2 S_3}{2\mathcal{M}_{xz}^{(2)}} \int_0^L \left(\frac{\partial C_L(\ell)}{\partial \dot{w}}\right)_0^e \varphi_{xz}^{(2)} d\ell + \frac{M_R}{\mathcal{M}_{xz}^{(2)}} \varphi_{xz}^{(2)}(\ell_R) \right\} s$$

$$+ \left\{ \frac{\rho U_0^2 S_3}{2\mathcal{M}_{xz}^{(2)}} \int_0^L \left[\left(\frac{\partial C_L(\ell)}{\partial w}\right)_0^e + \frac{1}{U_0} C_D(\ell)_0^e \right] \varphi_{xz}^{(2)} d\ell \right\}.$$

$$A_{2,19} = \left\{ \frac{\rho U_0^2 S_3}{2\mathcal{M}_{xz}^{(2)}} \int_0^L \left(\frac{\partial C_L(\ell)}{\partial p}\right)_0^e \varphi_{xz}^{(2)} d\ell \right\} s.$$

$$A_{2,20} = \left\{ \frac{M_R}{\mathcal{M}_{xz}^{(2)}} (L - \ell_{CG}) \varphi_{xz}^{(2)}(\ell_R) \right\} s^2 + \left\{ \frac{\rho U_0^2 S_3}{2\mathcal{M}_{xz}^{(2)}} \int_0^L \left(\frac{\partial C_L(\ell)}{\partial q}\right)_0^e \varphi_{xz}^{(2)} d\ell \right\} s.$$

$$A_{2,21} = \left\{ \frac{\rho U_0^2 S_3}{2\mathcal{M}_{xz}^{(2)}} \int_0^L \left(\frac{\partial C_L(\ell)}{\partial r}\right)_0^e \varphi_{xz}^{(2)} d\ell \right\} s.$$

$$A_{2,22} = \left\{ -\frac{M_{C_1}}{\mathcal{M}_{xz}^{(2)}} \sigma_{xz}^{(2)}(X_{1B}) \right\} s^2 + \left\{ \frac{K_1}{\mathcal{M}_{xz}^{(2)}} \varphi_{xz}^{(2)}(X_1) - \frac{M_{I_1}(T_s + T_c - D)}{M_t \mathcal{M}_{xz}^{(2)}} \sigma_{xz}^{(2)}(X_{1B}) \right\}.$$

$$A_{2,23} = \left\{ -\frac{M_{C_2}}{\mathcal{M}_{xz}^{(2)}} \sigma_{xz}^{(2)}(X_{2B}) \right\} s^2 + \left\{ \frac{K_2}{\mathcal{M}_{xz}^{(2)}} \varphi_{xz}^{(2)}(X_2) - \frac{M_{I_2}(T_s + T_c - D)}{M_t \mathcal{M}_{xz}^{(2)}} \sigma_{xz}^{(2)}(X_{2B}) \right\}.$$

$$A_{2,24} = \left\{ -\frac{M_{C_3}}{\mathcal{M}_{xz}^{(2)}} \sigma_{xz}^{(2)}(X_{3B}) \right\} s^2 + \left\{ \frac{K_3}{\mathcal{M}_{xz}^{(2)}} \varphi_{xz}^{(2)}(X_3) - \frac{M_{I_3}(T_s + T_c - D)}{M_t \mathcal{M}_{xz}^{(2)}} \sigma_{xz}^{(2)}(X_{3B}) \right\}.$$

$$A_{2,25} = \left\{ -\frac{M_{C_4}}{\mathcal{M}_{xz}^{(2)}} \sigma_{xz}^{(2)}(X_{4B}) \right\} s^2 + \left\{ \frac{K_4}{\mathcal{M}_{xz}^{(2)}} \varphi_{xz}^{(2)}(X_4) - \frac{M_{I_4}(T_s + T_c - D)}{M_t \mathcal{M}_{xz}^{(2)}} \sigma_{xz}^{(2)}(X_{4B}) \right\}.$$

$$A_{2,26} = A_{2,27} = A_{2,28} = A_{2,29} = 0.$$

$$A_{2,30} = \left\{ \frac{M_R \, \ell_R}{\mathcal{M}_{xz}^{(2)}} \, \varphi_{xz}^{(2)}(\ell_R) \right\} s^2 + \left\{ -\frac{T_c}{\mathcal{M}_{xz}^{(2)}} \, \varphi_{xz}^{(2)}(eG) \right\} .$$

$$A_{2,31} \; \cdots \cdots \; A_{2,40} = 0.$$

$$A_{3,1} = \left\{ \frac{M_R}{\mathcal{M}_{xz}^{(3)}} \left(\varphi_{xz}^{(1)}(eG) - \ell_R \, \sigma_{xz}^{(1)}(eG) \right) \varphi_{xz}^{(3)}(\ell_R) \right\} s^2 + \left\{ \frac{\rho U_0^2 S_3}{2\mathcal{M}_{xz}^{(3)}} \int_0^L \left[\left(\frac{\partial C_L(\ell)}{\partial w} \right)_0^e \right.\right.$$

$$\left.\left. + \frac{1}{U_0} C_D(\ell)_0^e \right] \varphi_{xz}^{(1)} \varphi_{xz}^{(3)} d\ell \right\} s + \left\{ - \frac{(T_s + T_c)}{\mathcal{M}_{xz}^{(3)}} \, \sigma_{xz}^{(1)}(eG) \, \varphi_{xz}^{(3)}(eG) \right.$$

$$\left. - \frac{\rho U_0^3 S_3}{2\mathcal{M}_{xz}^{(3)}} \int_0^L \left[\left(\frac{\partial C_L(\ell)}{\partial w} \right)_0^e + \frac{1}{U_0} C_D(\ell)_0^e \right] \sigma_{xz}^{(1)} \, \varphi_{xz}^{(3)} \, d\ell \right\} .$$

$$A_{3,2} = \left\{ \frac{M_R}{\mathcal{M}_{xz}^{(3)}} \left(\varphi_{xz}^{(2)}(eG) - \ell_R \, \sigma_{xz}^{(2)}(eG) \right) \varphi_{xz}^{(3)}(\ell_R) \right\} s^2 + \left\{ \frac{\rho U_0^2 S_3}{2\mathcal{M}_{xz}^{(3)}} \int_0^L \left[\left(\frac{\partial C_L(\ell)}{\partial w} \right)_0^e \right.\right.$$

$$\left.\left. + \frac{1}{U_0} C_D(\ell)_0^e \right] \varphi_{xz}^{(2)} \varphi_{xz}^{(3)} d\ell \right\} s + \left\{ - \frac{(T_s + T_c)}{\mathcal{M}_{xz}^{(3)}} \, \sigma_{xz}^{(2)}(eG) \, \varphi_{xz}^{(3)}(eG) \right.$$

$$\left. - \frac{\rho U_0^3 S_3}{2\mathcal{M}_{xz}^{(3)}} \int_0^L \left[\left(\frac{\partial C_L(\ell)}{\partial w} \right)_0^e + \frac{1}{U_0} C_D(\ell)_0^e \right] \sigma_{xz}^{(2)} \, \varphi_{xz}^{(3)} \, d\ell \right\} .$$

$$A_{3,3} = \left\{ 1 + \frac{M_R}{\mathcal{M}_{xz}^{(3)}} \left(\varphi_{xz}^{(3)}(eG) - \ell_R \, \sigma_{xz}^{(3)}(eG) \right) \varphi_{xz}^{(3)}(\ell_R) \right\} s^2 \left\{ 2 \xi_{xz}^{(3)} \omega_{xz}^{(3)} \right.$$

$$\left. + \frac{\rho U_0^2 S_3}{2\mathcal{M}_{xz}^{(3)}} \int_0^L \left[\left(\frac{\partial C_L(\ell)}{\partial w} \right)_0^e + \frac{1}{U_0} C_D(\ell)_0^e \right] \varphi_{xz}^{(3)2} d\ell \right\} s + \left\{ \omega_{xz}^{(3)2} \right.$$

$$\left. - \frac{(T_s + T_c)}{\mathcal{M}_{xz}^{(3)}} \, \sigma_{xz}^{(3)}(eG) \, \varphi_{xz}^{(3)}(eG) - \frac{\rho U_0^3 S_3}{2\mathcal{M}_{xz}^{(3)}} \int_0^L \left[\left(\frac{\partial C_L(\ell)}{\partial w} \right)_0^e + \frac{1}{U_0} C_D(\ell)_0^e \right] \sigma_{xz}^{(3)} \varphi_{xz}^{(3)} \, d\ell \right\} .$$

$$A_{3,4} = \left\{ \frac{M_R}{\mathcal{M}_{xz}^{(3)}} \left(\varphi_{xz}^{(4)}(eG) - \ell_R \, \sigma_{xz}^{(4)}(eG) \right) \varphi_{xz}^{(3)}(\ell_R) \right\} s^2 + \left\{ \frac{\rho U_0^2 S_3}{2\mathcal{M}_{xz}^{(3)}} \int_0^L \left[\left(\frac{\partial C_L(\ell)}{\partial w} \right)_0^e \right.\right.$$

$$\left.\left. + \frac{1}{U_0} C_D(\ell)_0^e \right] \varphi_{xz}^{(4)} \varphi_{xz}^{(3)} d\ell \right\} s + \left\{ - \frac{(T_s + T_c)}{\mathcal{M}_{xz}^{(3)}} \sigma_{xz}^{(4)}(eG) \, \varphi_{xz}^{(3)}(eG) \right.$$

$$\left. - \frac{\rho U_0^3 S_3}{2\mathcal{M}_{xz}^{(3)}} \int_0^L \left[\left(\frac{\partial C_L(\ell)}{\partial w} \right)_0^e + \frac{1}{U_0} C_D(\ell)_0^e \right] \sigma_{xz}^{(4)} \varphi_{xz}^{(3)} d\ell \right\} .$$

$$A_{3,5} = \left\{ \frac{M_R}{\mathcal{M}_{xz}^{(3)}} \left(\varphi_{xz}^{(5)}(eG) - \ell_R \, \sigma_{xz}^{(5)}(eG) \right) \varphi_{xz}^{(3)}(\ell_R) \right\} s^2 + \left\{ \frac{\rho U_0^2 S_3}{2\mathcal{M}_{xz}^{(3)}} \int_0^L \left[\left(\frac{\partial C_L(\ell)}{\partial w} \right)_0^e \right.\right.$$

$$\left.\left. + \frac{1}{U_0} C_D(\ell)_0^e \right] \varphi_{xz}^{(5)} \varphi_{xz}^{(3)} d\ell \right\} s + \left\{ - \frac{(T_s + T_c)}{\mathcal{M}_{xz}^{(3)}} \sigma_{xz}^{(5)}(eG) \, \varphi_{xz}^{(3)}(eG) \right.$$

$$\left. - \frac{\rho U_0^3 S_3}{2\mathcal{M}_{xz}^{(3)}} \int_0^L \left[\left(\frac{\partial C_L(\ell)}{\partial w} \right)_0^e + \frac{1}{U_0} C_D(\ell)_0^e \right] \sigma_{xz}^{(5)} \varphi_{xz}^{(3)} d\ell \right\} .$$

$$A_{3,6} = \left\{ \frac{\rho U_0^2 S_3}{2\mathcal{M}_{xz}^{(3)}} \int_0^L \left(\frac{\partial C_L(\ell)}{\partial v} \right)_0^e \varphi_{xy}^{(1)} \varphi_{xz}^{(3)} d\ell \right\} s$$

$$+ \left\{ - \frac{\rho U_0^3 S_3}{2\mathcal{M}_{xz}^{(3)}} \int_0^L \left(\frac{\partial C_L(\ell)}{\partial v} \right)_0^e \sigma_{xy}^{(1)} \varphi_{xz}^{(3)} d\ell \right\} .$$

$$A_{3,7} = \left\{ \frac{\rho U_0^2 S_3}{2\mathcal{M}_{xz}^{(3)}} \int_0^L \left(\frac{\partial C_L(\ell)}{\partial v} \right)_0^e \varphi_{xy}^{(2)} \varphi_{xz}^{(3)} d\ell \right\} s$$

$$+ \left\{ - \frac{\rho U_0^3 S_3}{2\mathcal{M}_{xz}^{(3)}} \int_0^L \left(\frac{\partial C_L(\ell)}{\partial v} \right)_0^e \sigma_{xy}^{(2)} \varphi_{xz}^{(3)} d\ell \right\} .$$

$$A_{3,8} = \left\{ \frac{\rho U_0^2 S_3}{2\mathcal{M}_{xz}^{(3)}} \int_0^L \left(\frac{\partial C_L(\ell)}{\partial v} \right)_0^e \varphi_{xy}^{(2)} \varphi_{xz}^{(3)} d\ell \right\} s$$

$$+ \left\{ - \frac{\rho U_0^3 S_3}{2\mathcal{M}_{xz}^{(3)}} \int_0^L \left(\frac{\partial C_L(\ell)}{\partial v} \right)_0^e \sigma_{xy}^{(3)} \varphi_{xz}^{(3)} d\ell \right\} .$$

$$A_{3,9} = \left\{ \frac{\rho U_0^2 S_3}{2 \mathcal{M}_{xz}^{(3)}} \int_0^L \left(\frac{\partial C_L(\ell)}{\partial v} \right)_0^e \varphi_{xy}^{(4)} \varphi_{xz}^{(3)} d\ell \right\} s$$

$$+ \left\{ - \frac{\rho U_0^3 S_3}{2 \mathcal{M}_{xz}^{(3)}} \int_0^L \left(\frac{\partial C_L(\ell)}{\partial v} \right)_0^e \sigma_{xy}^{(4)} \varphi_{xz}^{(3)} d\ell \right\} .$$

$$A_{3,10} = \left\{ \frac{\rho U_0^2 S_3}{2 \mathcal{M}_{xz}^{(3)}} \int_0^L \left(\frac{\partial C_L(\ell)}{\partial v} \right)_0^e \varphi_{xy}^{(5)} \varphi_{xz}^{(3)} d\ell \right\} s$$

$$+ \left\{ - \frac{\rho U_0^3 S_3}{2 \mathcal{M}_{xz}^{(3)}} \int_0^L \left(\frac{\partial C_L(\ell)}{\partial v} \right)_0^e \sigma_{xy}^{(5)} \varphi_{xz}^{(3)} d\ell \right\} .$$

$$A_{3,11} = \left\{ \frac{\rho U_0^2 S_3}{2 \mathcal{M}_{xz}^{(3)}} \int_0^L \left(\frac{\partial C_L(\ell)}{\partial p} \right)_0^e \varphi_{zy}^{(1)} \varphi_{xz}^{(3)} d\ell \right\} s .$$

$$A_{3,12} = \left\{ \frac{\rho U_0^2 S_3}{2 \mathcal{M}_{xz}^{(3)}} \int_0^L \left(\frac{\partial C_L(\ell)}{\partial p} \right)_0^e \varphi_{zy}^{(2)} \varphi_{xz}^{(3)} d\ell \right\} s .$$

$$A_{3,13} = \left\{ \frac{\rho U_0^2 S_3}{2 \mathcal{M}_{xz}^{(3)}} \int_0^L \left(\frac{\partial C_L(\ell)}{\partial p} \right)_0^e \varphi_{zy}^{(3)} \varphi_{xz}^{(3)} d\ell \right\} s .$$

$$A_{3,14} = \left\{ \frac{\rho U_0^2 S_3}{2 \mathcal{M}_{xz}^{(3)}} \int_0^L \left(\frac{\partial C_L(\ell)}{\partial p} \right)_0^e \varphi_{zy}^{(4)} \varphi_{xz}^{(3)} d\ell \right\} s .$$

$$A_{3,15} = \left\{ \frac{\rho U_0^2 S_3}{2 \mathcal{M}_{xz}^{(3)}} \int_0^L \left(\frac{\partial C_L(\ell)}{\partial p} \right)_0^e \varphi_{zy}^{(5)} \varphi_{xz}^{(3)} d\ell \right\} s .$$

$$A_{3,16} = \left\{ \frac{\rho U_0^2 S_3}{2 \mathcal{M}_{xz}^{(3)}} \int_0^L \left[\left(\frac{\partial C_L(\ell)}{\partial u} \right)_0^e + \frac{2}{U_0} C_L(\ell)_0^e \right] \varphi_{xz}^{(3)} d\ell \right\} .$$

$$A_{3,17} = \left\{ \frac{\rho U_0^2 S_3}{2 \mathcal{M}_{xz}^{(3)}} \int_0^L \left(\frac{\partial C_L(\ell)}{\partial v} \right)_0^e \varphi_{xz}^{(3)} d\ell \right\} .$$

$$A_{3,18} = \left\{ \frac{\rho U_0^2 S_3}{2\mathcal{M}_{xz}^{(3)}} \int_0^L \left(\frac{\partial C_L(\ell)}{\partial \dot{w}} \right)_0^e \varphi_{xz}^{(3)} d\ell + \frac{M_R}{\mathcal{M}_{xz}^{(3)}} \varphi_{xz}^{(3)}(\ell_R) \right\} s$$

$$+ \left\{ \frac{\rho U_0^2 S_3}{2\mathcal{M}_{xz}^{(3)}} \int_0^L \left[\left(\frac{\partial C_L(\ell)}{\partial w} \right)_0^e + \frac{1}{U_0} C_D (\ell)_0^e \right] \varphi_{xz}^{(3)} d\ell \right\}.$$

$$A_{3,19} = \left\{ \frac{\rho U_0^2 S_3}{2\mathcal{M}_{xz}^{(3)}} \int_0^L \left(\frac{\partial C_L(\ell)}{\partial p} \right)_0^e \varphi_{xz}^{(3)} d\ell \right\} s$$

$$A_{3,20} = \left\{ \frac{M_R}{\mathcal{M}_{xz}^{(3)}} (L - \ell_{CG}) \varphi_{xz}^{(3)}(\ell_R) \right\} s^2 + \left\{ \frac{\rho U_0^2 S_3}{2\mathcal{M}_{xz}^{(3)}} \int_0^L \left(\frac{\partial C_L(\ell)}{\partial q} \right)_0^e \varphi_{xz}^{(3)} d\ell \right\} s$$

$$A_{3,21} = \left\{ \frac{\rho U_0^2 S_3}{2\mathcal{M}_{xz}^{(3)}} \int_0^L \left(\frac{\partial C_L(\ell)}{\partial r} \right)_0^e \varphi_{xz}^{(3)} d\ell \right\} s.$$

$$A_{3,22} = \left\{ -\frac{M_{C_1}}{\mathcal{M}_{xz}^{(3)}} \sigma_{xz}^{(3)}(X_{1B}) \right\} s^2 + \left\{ \frac{K_1}{\mathcal{M}_{xz}^{(3)}} \varphi_{xz}^{(3)}(X_1) - \frac{M_{l_1}(T_s + T_c - D)}{M_t \mathcal{M}_{xz}^{(3)}} \sigma_{xz}^{(3)}(X_{1B}) \right\}$$

$$A_{3,23} = \left\{ -\frac{M_{C_2}}{\mathcal{M}_{xz}^{(3)}} \sigma_{xz}^{(3)}(X_{2B}) \right\} s^2 + \left\{ \frac{K_2}{\mathcal{M}_{xz}^{(3)}} \varphi_{xz}^{(3)}(X_2) - \frac{M_{l_2}(T_s + T_c - D)}{M_t \mathcal{M}_{xz}^{(3)}} \sigma_{xz}^{(3)}(X_{2B}) \right\}$$

$$A_{3,24} = \left\{ -\frac{M_{C_3}}{\mathcal{M}_{xz}^{(3)}} \sigma_{xz}^{(3)}(X_{3B}) \right\} s^2 + \left\{ \frac{K_3}{\mathcal{M}_{xz}^{(3)}} \varphi_{xz}^{(3)}(X_3) - \frac{M_{l_3}(T_s + T_c - D)}{M_t \mathcal{M}_{xz}^{(3)}} \sigma_{xz}^{(3)}(X_{3B}) \right\}$$

$$A_{3,25} = \left\{ -\frac{M_{C_4}}{\mathcal{M}_{xz}^{(3)}} \sigma_{xz}^{(3)}(X_{4B}) \right\} s^2 + \left\{ \frac{K_4}{\mathcal{M}_{xz}^{(3)}} \varphi_{xz}^{(3)}(X_4) - \frac{M_{l_4}(T_s + T_c - D)}{M_t \mathcal{M}_{xz}^{(3)}} \sigma_{xz}^{(3)}(X_{4B}) \right\}$$

$$A_{3,26} \cdots A_{3,29} = 0.$$

$$A_{3,30} = \left\{ \frac{M_R \ell_R}{\mathcal{M}_{xz}^{(3)}} \varphi_{xz}^{(3)}(\ell_R) \right\} s^2 + \left\{ \frac{T_C}{\mathcal{M}_{xz}^{(3)}} \varphi_{xz}^{(3)}(eG) \right\}.$$

$$A_{3,31} \cdots A_{3,40} = 0$$

$$A_{4,1} = \left\{ \frac{M_R}{\mathcal{M}_{xz}^{(4)}} \left(\varphi_{xz}^{(1)}(eG) - \ell_R \, \sigma_{xz}^{(1)}(eG) \right) \varphi_{xz}^{(4)}(\ell_R) \right\} s^2 + \left\{ \frac{\rho U_0^2 S_3}{2\mathcal{M}_{xz}^{(4)}} \int_0^L \left[\left(\frac{\partial C_L(\ell)}{\partial w} \right)_0^e \right. \right.$$

$$\left. \left. + \frac{1}{U_0} C_D(\ell)_0^e \right] \varphi_{xz}^{(1)} \varphi_{xz}^{(4)} d\ell \right\} s + \left\{ - \frac{(T_s + T_c)}{\mathcal{M}_{xz}^{(4)}} \sigma_{xz}^{(1)}(eG) \, \varphi_{xz}^{(4)}(eG) \right.$$

$$\left. - \frac{\rho U_0^3 S_3}{2\mathcal{M}_{xz}^{(4)}} \int_0^L \left[\left(\frac{\partial C_L(\ell)}{\partial w} \right)_0^e + \frac{1}{U_0} C_D(\ell)_0^e \right] \sigma_{xz}^{(1)} \varphi_{xz}^{(4)} d\ell \right\} .$$

$$A_{4,2} = \left\{ \frac{M_R}{\mathcal{M}_{xz}^{(4)}} \left(\varphi_{xz}^{(2)}(eG) - \ell_R \, \sigma_{xz}^{(2)}(eG) \right) \varphi_{xz}^{(4)}(\ell_R) \right\} s^2 + \left\{ \frac{\rho U_0^2 S_3}{2\mathcal{M}_{xz}^{(4)}} \int_0^L \left[\left(\frac{\partial C_L(\ell)}{\partial w} \right)_0^e \right. \right.$$

$$\left. \left. + \frac{1}{U_0} C_D(\ell)_0^e \right] \varphi_{xz}^{(2)} \varphi_{xz}^{(4)} d\ell \right\} s + \left\{ - \frac{(T_s + T_c)}{\mathcal{M}_{xz}^{(4)}} \sigma_{xz}^{(2)}(eG) \, \varphi_{xz}^{(4)}(eG) \right.$$

$$\left. - \frac{\rho U_0^3 S_3}{2\mathcal{M}_{xz}^{(4)}} \int_0^L \left[\left(\frac{\partial C_L(\ell)}{\partial w} \right)_0^e + \frac{1}{U_0} C_D(\ell)_0^e \right] \sigma_{xz}^{(2)} \varphi_{xz}^{(4)} d\ell \right\} .$$

$$A_{4,3} = \left\{ \frac{M_R}{\mathcal{M}_{xz}^{(4)}} \left(\varphi_{xz}^{(3)}(eG) - \ell_R \, \sigma_{xz}^{(3)}(eG) \right) \varphi_{xz}^{(4)}(\ell_R) \right\} s^2 + \left\{ \frac{\rho U_0^2 S_3}{2\mathcal{M}_{xz}^{(4)}} \int_0^L \left[\left(\frac{\partial C_L(\ell)}{\partial w} \right)_0^e \right. \right.$$

$$\left. \left. + \frac{1}{U_0} C_D(\ell)_0^e \right] \varphi_{xz}^{(3)} \varphi_{xz}^{(4)} d\ell \right\} s + \left\{ - \frac{(T_s + T_c)}{\mathcal{M}_{xz}^{(4)}} \sigma_{xz}^{(3)}(eG) \, \varphi_{xz}^{(4)}(eG) \right.$$

$$\left. - \frac{\rho U_0^3 S_3}{2\mathcal{M}_{xz}^{(4)}} \int_0^L \left[\left(\frac{\partial C_L(\ell)}{\partial w} \right)_0^e + \frac{1}{U_0} C_D(\ell)_0^e \right] \sigma_{xz}^{(3)} \varphi_{xz}^{(4)} d\ell \right\} .$$

$$A_{4,4} = \left\{ 1 + \frac{M_R}{\mathcal{M}_{xz}^{(4)}} \left(\varphi_{xz}^{(4)}(eG) - \ell_R \, \sigma_{xz}^{(4)}(eG) \right) \varphi_{xz}^{(4)}(\ell_R) \right\} s^2 + \left\{ 2 \xi_{xz}^{(4)} \omega_{xz}^{(4)} \right.$$

$$\left. + \frac{\rho U_0^2 S_3}{2\mathcal{M}_{xz}^{(4)}} \int_0^L \left[\left(\frac{\partial C_L(\ell)}{\partial w} \right)_0^e + \frac{1}{U_0} C_D(\ell)_0^e \right] \varphi_{xz}^{(4)2} d\ell \right\} s + \left\{ \omega_{xz}^{(4)2} \right.$$

$$\left. - \frac{(T_s + T_c)}{\mathcal{M}_{xz}^{(4)}} \sigma_{xz}^{(4)}(eG) \, \varphi_{xz}^{(4)}(eG) - \frac{\rho U_0^3 S_3}{2\mathcal{M}_{xz}^{(4)}} \int_0^L \left[\left(\frac{\partial C_L(\ell)}{\partial w} \right)_0^e + \frac{1}{U_0} C_D(\ell)_0^e \right] \sigma_{xz}^{(4)} \varphi_{xz}^{(4)} d\ell \right\} .$$

$$A_{4,5} = \left\{ \frac{M_R}{\mathcal{M}_{xz}^{(4)}} \left(\varphi_{xz}^{(5)}(eG) - \ell_R\, \sigma_{xz}^{(5)}(eG) \right) \varphi_{xz}^{(4)}(\ell_R) \right\} s^2 + \left\{ \frac{\rho U_0^2 S_3}{2\mathcal{M}_{xz}^{(4)}} \int_0^L \left[\left(\frac{\partial C_L(\ell)}{\partial w} \right)_0^e \right.\right.$$

$$\left.\left. + \frac{1}{U_0} C_D(\ell)_0^e \right] \varphi_{xz}^{(5)} \varphi_{xz}^{(4)} d\ell \right\} s + \left\{ - \frac{(T_s + T_c)}{\mathcal{M}_{xz}^{(4)}} \sigma_{xz}^{(5)}(eG)\, \varphi_{xz}^{(4)}(eG) \right.$$

$$\left. - \frac{\rho U_0^3 S_3}{2\mathcal{M}_{xz}^{(4)}} \int_0^L \left[\left(\frac{\partial C_L(\ell)}{\partial w} \right)_0^e + \frac{1}{U_0} C_D(\ell)_0^e \right] \sigma_{xz}^{(5)} \varphi_{xz}^{(4)} d\ell \right\}.$$

$$A_{4,6} = \left\{ \frac{\rho U_0^2 S_3}{2\mathcal{M}_{xz}^{(4)}} \int_0^L \left(\frac{\partial C_L(\ell)}{\partial v} \right)_0^e \varphi_{xy}^{(1)} \varphi_{xz}^{(4)} d\ell \right\} s$$

$$+ \left\{ - \frac{\rho U_0^3 S_3}{2\mathcal{M}_{xz}^{(4)}} \int_0^L \left(\frac{\partial C_L(\ell)}{\partial v} \right)_0^e \sigma_{xy}^{(1)} \varphi_{xz}^{(4)} d\ell \right\},$$

$$A_{4,7} = \left\{ \frac{\rho U_0^2 S_3}{2\mathcal{M}_{xz}^{(4)}} \int_0^L \left(\frac{\partial C_L(\ell)}{\partial v} \right)_0^e \varphi_{xy}^{(2)} \varphi_{xz}^{(4)} d\ell \right\} s$$

$$+ \left\{ - \frac{\rho U_0^3 S_3}{2\mathcal{M}_{xz}^{(4)}} \int_0^L \left(\frac{\partial C_L(\ell)}{\partial v} \right)_0^e \sigma_{xy}^{(2)} \varphi_{xz}^{(4)} d\ell \right\},$$

$$A_{4,8} = \left\{ \frac{\rho U_0^2 S_3}{2\mathcal{M}_{xz}^{(4)}} \int_0^L \left(\frac{\partial C_L(\ell)}{\partial v} \right)_0^e \varphi_{xy}^{(3)} \varphi_{xz}^{(4)} d\ell \right\} s$$

$$+ \left\{ - \frac{\rho U_0^3 S_3}{2\mathcal{M}_{xz}^{(4)}} \int_0^L \left(\frac{\partial C_L(\ell)}{\partial v} \right)_0^e \sigma_{xy}^{(3)} \varphi_{xz}^{(4)} d\ell \right\},$$

$$A_{4,9} = \left\{ \frac{\rho U_0^2 S_3}{2\mathcal{M}_{xz}^{(4)}} \int_0^L \left(\frac{\partial C_L(\ell)}{\partial v} \right)_0^e \varphi_{xy}^{(4)} \varphi_{xz}^{(4)} d\ell \right\} s$$

$$+ \left\{ - \frac{\rho U_0^3 S_3}{2\mathcal{M}_{xz}^{(4)}} \int_0^L \left(\frac{\partial C_L(\ell)}{\partial v} \right)_0^e \sigma_{xy}^{(4)} \varphi_{xz}^{(4)} d\ell \right\},$$

$$A_{4,10} = \left\{ \frac{\rho U_0^2 S_3}{2\mathcal{M}_{xz}^{(4)}} \int_0^L \left(\frac{\partial C_L(\ell)}{\partial v} \right)_0^e \varphi_{xy}^{(5)} \varphi_{xz}^{(4)} d\ell \right\} s$$

$$+ \left\{ - \frac{\rho U_0^3 S_3}{2\mathcal{M}_{xz}^{(4)}} \int_0^L \left(\frac{\partial C_L(\ell)}{\partial v} \right)_0^e \sigma_{xy}^{(5)} \varphi_{xz}^{(4)} d\ell \right\},$$

$$A_{4,11} = \left\{ \frac{\rho U_0^2 S_3}{2 \mathcal{M}_{xz}^{(4)}} \int_0^L \left(\frac{\partial C_L(\ell)}{\partial p} \right)_0^e \varphi_{zy}^{(1)} \varphi_{xz}^{(4)} d\ell \right\} s.$$

$$A_{4,12} = \left\{ \frac{\rho U_0^2 S_3}{2 \mathcal{M}_{xz}^{(4)}} \int_0^L \left(\frac{\partial C_L(\ell)}{\partial p} \right)_0^e \varphi_{zy}^{(2)} \varphi_{xz}^{(4)} d\ell \right\} s.$$

$$A_{4,13} = \left\{ \frac{\rho U_0^2 S_3}{2 \mathcal{M}_{xz}^{(4)}} \int_0^L \left(\frac{\partial C_L(\ell)}{\partial p} \right)_0^e \varphi_{zy}^{(3)} \varphi_{xz}^{(4)} d\ell \right\} s.$$

$$A_{4,14} = \left\{ \frac{\rho U_0^2 S_3}{2 \mathcal{M}_{xz}^{(4)}} \int_0^L \left(\frac{\partial C_L(\ell)}{\partial p} \right)_0^e \varphi_{zy}^{(4)} \varphi_{xz}^{(4)} d\ell \right\} s.$$

$$A_{4,15} = \left\{ \frac{\rho U_0^2 S_3}{2 \mathcal{M}_{xz}^{(4)}} \int_0^L \left(\frac{\partial C_L(\ell)}{\partial p} \right)_0^e \varphi_{zy}^{(5)} \varphi_{xz}^{(4)} d\ell \right\} s.$$

$$A_{4,16} = \left\{ \frac{\rho U_0^2 S_3}{2 \mathcal{M}_{xz}^{(4)}} \int_0^L \left[\left(\frac{\partial C_L(\ell)}{\partial u} \right)_0^e + \frac{2}{U_0} C_L(\ell)_0^e \right] \varphi_{xz}^{(4)} d\ell \right\}.$$

$$A_{4,17} = \left\{ \frac{\rho U_0^2 S_3}{2 \mathcal{M}_{xz}^{(4)}} \int_0^L \left(\frac{\partial C_L(\ell)}{\partial v} \right)_0^e \varphi_{xz}^{(4)} d\ell \right\}.$$

$$A_{4,18} = \left\{ \frac{\rho U_0^2 S_3}{2 \mathcal{M}_{xz}^{(4)}} \int_0^L \left(\frac{\partial C_L(\ell)}{\partial \dot{w}} \right)_0^e \varphi_{xz}^{(4)} d\ell + \frac{M_R}{\mathcal{M}_{xz}^{(4)}} \varphi_{xz}^{(4)}(\ell_R) \right\} s$$

$$+ \left\{ \frac{\rho U_0^2 S_3}{2 \mathcal{M}_{xz}^{(4)}} \int_0^L \left[\left(\frac{\partial C_L(\ell)}{\partial w} \right)_0^e + \frac{1}{U_0} C_D(\ell)_0^e \right] \varphi_{xz}^{(4)} d\ell \right\}.$$

$$A_{4,19} = \left\{ \frac{\rho U_0^2 S_3}{2 \mathcal{M}_{xz}^{(4)}} \int_0^L \left(\frac{\partial C_L(\ell)}{\partial p} \right)_0^e \varphi_{xz}^{(4)} d\ell \right\} s$$

$$A_{4,20} = \left\{ \frac{M_R}{\mathcal{M}_{xz}^{(4)}} (L - \ell_{CG}) \varphi_{xz}^{(4)}(\ell_R) \right\} s^2 + \left\{ \frac{\rho U_0^2 S_3}{2 \mathcal{M}_{xz}^{(4)}} \int_0^L \left(\frac{\partial C_L(\ell)}{\partial q} \right)_0^e \varphi_{xz}^{(4)} d\ell \right\} s.$$

$$A_{4,21} = \left\{ \frac{\rho U_0^2 S_3}{2 \mathcal{M}_{xz}^{(4)}} \int_0^L \left(\frac{\partial C_L(\ell)}{\partial r} \right)_0^e \varphi_{xz}^{(4)} \, d\ell \right\} s .$$

$$A_{4,22} = \left\{ - \frac{M_{C_1}}{\mathcal{M}_{xz}^{(4)}} \sigma_{xz}^{(4)} (X_{1B}) \right\} s^2 + \left\{ \frac{K_1}{\mathcal{M}_{xz}^{(4)}} \varphi_{xz}^{(4)} (X_1) - \frac{M_{I_1} (T_s + T_c - D)}{M_t \mathcal{M}_{xz}^{(4)}} \sigma_{xz}^{(4)} (X_{1B}) \right\}$$

$$A_{4,23} = \left\{ \frac{-M_{C_2}}{\mathcal{M}_{xz}^{(4)}} \sigma_{xz}^{(4)} (X_{2B}) \right\} s^2 + \left\{ \frac{K_2}{\mathcal{M}_{xz}^{(4)}} \varphi_{xz}^{(4)} (X_2) - \frac{M_{I_2} (T_s + T_c - D)}{M_t \mathcal{M}_{xz}^{(4)}} \sigma_{xz}^{(4)} (X_{2B}) \right\}$$

$$A_{4,24} = \left\{ \frac{-M_{C_3}}{\mathcal{M}_{xz}^{(4)}} \sigma_{xz}^{(4)} (X_{3B}) \right\} s^2 + \left\{ \frac{K_3}{\mathcal{M}_{xz}^{(4)}} \varphi_{xz}^{(4)} (X_3) - \frac{M_{I_3} (T_s + T_c - D)}{M_t \mathcal{M}_{xz}^{(4)}} \sigma_{xz}^{(4)} (X_{3B}) \right\}$$

$$A_{4,25} = \left\{ - \frac{M_{C_4}}{\mathcal{M}_{xz}^{(4)}} \sigma_{xz}^{(4)} (X_{4B}) \right\} s^2 + \left\{ \frac{K_4}{\mathcal{M}_{xz}^{(4)}} \varphi_{xz}^{(4)} (X_4) - \frac{M_{I_4} (T_s + T_c - D)}{M_t \mathcal{M}_{xz}^{(4)}} \sigma_{xz}^{(4)} (X_{4B}) \right\}$$

$$A_{4,26} \cdots\cdots A_{4,29} = 0.$$

$$A_{4,30} = \left\{ \frac{M_R \, \ell_R}{\mathcal{M}_{xz}^{(4)}} \varphi_{xz}^{(4)} (\ell_R) \right\} s^2 + \left\{ \frac{T_c}{\mathcal{M}_{xz}^{(4)}} \varphi_{xz}^{(4)} (eG) \right\} .$$

$$A_{4,31} \cdots\cdots A_{4,40} = 0 .$$

$$A_{5,1} = \left\{ \frac{M_R}{\mathcal{M}_{xz}^{(5)}} \left(\varphi_{xz}^{(1)} (eG) - \ell_R \sigma_{xz}^{(1)} (eG) \right) \varphi_{xz}^{(5)} (\ell_R) \right\} s^2 + \left\{ \frac{\rho U_0^2 S_3}{2 \mathcal{M}_{xz}^{(5)}} \int_0^L \left[\left(\frac{\partial C_L(\ell)}{\partial w} \right)_0^e \right.\right.$$

$$\left.\left. + \frac{1}{U_0} C_D(\ell)_0^e \right] \varphi_{xz}^{(1)} \varphi_{xz}^{(5)} d\ell \right\} s + \left\{ - \frac{(T_s + T_c)}{\mathcal{M}_{xz}^{(5)}} \sigma_{xz}^{(1)} (eG) \, \varphi_{xz}^{(5)} (eG) \right.$$

$$\left. - \frac{\rho U_0^3 S_3}{2 \mathcal{M}_{xz}^{(5)}} \int_0^L \left[\left(\frac{\partial C_L(\ell)}{\partial w} \right)_0^e + \frac{1}{U_0} C_D(\ell)_0^e \right] \sigma_{xz}^{(1)} \varphi_{xz}^{(5)} d\ell \right\} .$$

$$A_{5,2} = \left\{ \frac{M_R}{\mathcal{M}_{xz}^{(5)}} \left(\varphi_{xz}^{(2)}(eG) - \ell_R \, \sigma_{xz}^{(2)}(eG) \right) \varphi_{xz}^{(5)}(\ell_R) \right\} s^2 + \left\{ \frac{\rho U_0^2 S_3}{2\mathcal{M}_{xz}^{(5)}} \int_0^L \left[\left(\frac{\partial C_L(\ell)}{\partial w} \right)_0^e \right. \right.$$

$$\left. \left. + \frac{1}{U_0} C_D(\ell)_0^e \right] \varphi_{xz}^{(2)} \varphi_{xz}^{(5)} d\ell \right\} s + \left\{ - \frac{(T_s + T_c)}{\mathcal{M}_{xz}^{(5)}} \sigma_{xz}^{(2)}(eG) \, \varphi_{xz}^{(5)}(eG) \right.$$

$$\left. - \frac{\rho U_0^3 S_3}{2\mathcal{M}_{xz}^{(5)}} \int_0^L \left[\left(\frac{\partial C_L(\ell)}{\partial w} \right)_0^e + \frac{1}{U_0} C_D(\ell)_0^e \right] \sigma_{xz}^{(2)} \varphi_{xz}^{(5)} d\ell \right\} .$$

$$A_{5,3} = \left\{ \frac{M_R}{\mathcal{M}_{xz}^{(5)}} \left(\varphi_{xz}^{(3)}(eG) - \ell_R \, \sigma_{xz}^{(3)}(eG) \right) \varphi_{xz}^{(5)}(\ell_R) \right\} s^2 + \left\{ \frac{\rho U_0^2 S_3}{2\mathcal{M}_{xz}^{(5)}} \int_0^L \left[\left(\frac{\partial C_L(\ell)}{\partial w} \right)_0^e \right. \right.$$

$$\left. \left. + \frac{1}{U_0} C_D(\ell)_0^e \right] \varphi_{xz}^{(3)} \varphi_{xz}^{(5)} d\ell \right\} s + \left\{ - \frac{(T_s + T_c)}{\mathcal{M}_{xz}^{(5)}} \sigma_{xz}^{(3)}(eG) \, \varphi_{xz}^{(5)}(eG) \right.$$

$$\left. - \frac{\rho U_0^3 S_3}{2\mathcal{M}_{xz}^{(5)}} \int_0^L \left[\left(\frac{\partial C_L(\ell)}{\partial w} \right)_0^e + \frac{1}{U_0} C_D(\ell)_0^e \right] \sigma_{xz}^{(3)} \varphi_{xz}^{(5)} d\ell \right\} .$$

$$A_{5,4} = \left\{ \frac{M_R}{\mathcal{M}_{xz}^{(5)}} \left(\varphi_{xz}^{(4)}(eG) - \ell_R \, \sigma_{xz}^{(4)}(eG) \right) \varphi_{xz}^{(5)}(\ell_R) \right\} s^2 + \left\{ \frac{\rho U_0^2 S_3}{2\mathcal{M}_{xz}^{(5)}} \int_0^L \left[\left(\frac{\partial C_L(\ell)}{\partial w} \right)_0^e \right. \right.$$

$$\left. \left. + \frac{1}{U_0} C_D(\ell)_0^e \right] \varphi_{xz}^{(4)} \varphi_{xz}^{(5)} d\ell \right\} s + \left\{ - \frac{(T_s + T_c)}{\mathcal{M}_{xz}^{(5)}} \sigma_{xz}^{(4)}(eG) \, \varphi_{xz}^{(5)}(eG) \right.$$

$$\left. - \frac{\rho U_0^3 S_3}{2\mathcal{M}_{xz}^{(5)}} \int_0^L \left[\left(\frac{\partial C_L(\ell)}{\partial w} \right)_0^e + \frac{1}{U_0} C_D(\ell)_0^e \right] \sigma_{xz}^{(4)} \varphi_{xz}^{(5)} d\ell \right\} .$$

$$A_{5,5} = \left\{ 1 + \frac{M_R}{\mathcal{M}_{xz}^{(5)}} \left(\varphi_{xz}^{(5)}(eG) - \ell_R \, \sigma_{xz}^{(5)}(eG) \right) \varphi_{xz}^{(5)}(\ell_R) \right\} s^2 \left\{ 2 \xi_{xz}^{(5)} \omega_{xz}^{(5)} \right.$$

$$\left. + \frac{\rho U_0^2 S_3}{2\mathcal{M}_{xz}^{(5)}} \int_0^L \left[\left(\frac{\partial C_L(\ell)}{\partial w} \right)_0^e + \frac{1}{U_0} C_D(\ell)_0^e \right] \varphi_{xz}^{(5)2} d\ell \right\} s + \left\{ \omega_{xz}^{(5)2} \right.$$

$$\left. - \frac{(T_s + T_c)}{\mathcal{M}_{xz}^{(5)}} \sigma_{xz}^{(5)}(eG) \, \varphi_{xz}^{(5)}(eG) - \frac{\rho U_0^3 S_3}{2\mathcal{M}_{xz}^{(5)}} \int_0^L \left[\left(\frac{\partial C_L(\ell)}{\partial w} \right)_0^e + \frac{1}{U_0} C_D(\ell)_0^e \right] \sigma_{xz}^{(5)} \varphi_{xz}^{(5)} d\ell \right\} .$$

$$A_{5'6} = \left\{ \frac{\rho U_0^2 S_3}{2 \mathcal{M}_{xz}^{(5)}} \int_0^L \left(\frac{\partial C_L(\ell)}{\partial v} \right)_0^e \varphi_{xy}^{(1)} \varphi_{xz}^{(5)} d\ell \right\} s$$

$$+ \left\{ - \frac{\rho U_0^3 S_3}{2 \mathcal{M}_{xz}^{(5)}} \int_0^L \left(\frac{\partial C_L(\ell)}{\partial v} \right)_0^e \sigma_{xy}^{(1)} \varphi_{xz}^{(5)} d\ell \right\},$$

$$A_{5'7} = \left\{ \frac{\rho U_0^2 S_3}{2 \mathcal{M}_{xz}^{(5)}} \int_0^L \left(\frac{\partial C_L(\ell)}{\partial v} \right)_0^e \varphi_{xy}^{(2)} \varphi_{xz}^{(5)} d\ell \right\} s$$

$$+ \left\{ - \frac{\rho U_0^3 S_3}{2 \mathcal{M}_{xz}^{(5)}} \int_0^L \left(\frac{\partial C_L(\ell)}{\partial v} \right)_0^e \sigma_{xy}^{(2)} \varphi_{xz}^{(5)} d\ell \right\},$$

$$A_{5'8} = \left\{ \frac{\rho U_0^2 S_3}{2 \mathcal{M}_{xz}^{(5)}} \int_0^L \left(\frac{\partial C_L(\ell)}{\partial v} \right)_0^e \varphi_{xy}^{(3)} \varphi_{xz}^{(5)} d\ell \right\} s$$

$$+ \left\{ - \frac{\rho U_0^3 S_3}{2 \mathcal{M}_{xz}^{(5)}} \int_0^L \left(\frac{\partial C_L(\ell)}{\partial v} \right)_0^e \sigma_{xy}^{(3)} \varphi_{xz}^{(5)} d\ell \right\},$$

$$A_{5'9} = \left\{ \frac{\rho U_0^2 S_3}{2 \mathcal{M}_{xz}^{(5)}} \int_0^L \left(\frac{\partial C_L(\ell)}{\partial v} \right)_0^e \varphi_{xy}^{(4)} \varphi_{xz}^{(5)} d\ell \right\} s$$

$$+ \left\{ - \frac{\rho U_0^3 S_3}{2 \mathcal{M}_{xz}^{(5)}} \int_0^L \left(\frac{\partial C_L(\ell)}{\partial v} \right)_0^e \sigma_{xy}^{(4)} \varphi_{xz}^{(5)} d\ell \right\},$$

$$A_{5'10} = \left\{ \frac{\rho U_0^2 S_3}{2 \mathcal{M}_{xz}^{(5)}} \int_0^L \left(\frac{\partial C_L(\ell)}{\partial v} \right)_0^e \varphi_{xy}^{(5)} \varphi_{xz}^{(5)} d\ell \right\} s$$

$$+ \left\{ - \frac{\rho U_0^3 S_3}{2 \mathcal{M}_{xz}^{(5)}} \int_0^L \left(\frac{\partial C_L(\ell)}{\partial v} \right)_0^e \sigma_{xy}^{(5)} \varphi_{xz}^{(5)} d\ell \right\},$$

$$A_{5'11} = \left\{ \frac{\rho U_0^2 S_3}{2 \mathcal{M}_{xz}^{(5)}} \int_0^L \left(\frac{\partial C_L(\ell)}{\partial p} \right)_0^e \varphi_{zy}^{(1)} \varphi_{xz}^{(5)} d\ell \right\} s.$$

$$A_{5'12} = \left\{ \frac{\rho U_0^2 S_3}{2 \mathcal{M}_{xz}^{(5)}} \int_0^L \left(\frac{\partial C_L(\ell)}{\partial p} \right)_0^e \varphi_{zy}^{(2)} \varphi_{xz}^{(5)} d\ell \right\} s.$$

$$A_{5'13} = \left\{ \frac{\rho U_0^2 S_3}{2 \mathcal{M}_{xz}^{(5)}} \int_0^L \left(\frac{\partial C_L(\ell)}{\partial p} \right)_0^e \varphi_{zy}^{(3)} \varphi_{xz}^{(5)} d\ell \right\} s.$$

$$A_{5'14} = \left\{ \frac{\rho U_0^2 S_3}{2 \mathcal{M}_{xz}^{(5)}} \int_0^L \left(\frac{\partial C_L(\ell)}{\partial p} \right)_0^e \varphi_{zy}^{(4)} \varphi_{xz}^{(5)} d\ell \right\} s.$$

$$A_{5'15} = \left\{ \frac{\rho U_0^2 S_3}{2 \mathcal{M}_{xz}^{(5)}} \int_0^L \left(\frac{\partial C_L(\ell)}{\partial p} \right)_0^e \varphi_{zy}^{(5)} \varphi_{xz}^{(5)} d\ell \right\} s.$$

$$A_{5'16} = \left\{ \frac{\rho U_0^2 S_3}{2 \mathcal{M}_{xz}^{(5)}} \int_0^L \left[\left(\frac{\partial C_L(\ell)}{\partial u} \right)_0^e + \frac{2}{U_0} C_L(\ell)_0^e \right] \varphi_{xz}^{(5)} d\ell \right\}.$$

$$A_{5'17} = \left\{ \frac{\rho U_0^2 S_5}{2 \mathcal{M}_{xz}^{(5)}} \int_0^L \left(\frac{\partial C_L(\ell)}{\partial v} \right)_0^e \varphi_{xz}^{(5)} d\ell \right\}.$$

$$A_{5'18} = \left\{ \frac{\rho U_0^2 S_3}{2 \mathcal{M}_{xz}^{(5)}} \int_0^L \left(\frac{\partial C_L(\ell)}{\partial \dot{w}} \right)_0^e \varphi_{xz}^{(5)} d\ell + \frac{M_R}{\mathcal{M}_{xz}^{(5)}} \varphi_{xz}^{(5)}(\ell_R) \right\} s$$

$$+ \left\{ \frac{\rho U_0^2 S_3}{2 \mathcal{M}_{xz}^{(5)}} \int_0^L \left[\left(\frac{\partial C_L(\ell)}{\partial w} \right)_0^e + \frac{1}{U_0} C_D(\ell)_0^e \right] \varphi_{xz}^{(5)} d\ell \right\}.$$

$$A_{5'19} = \left\{ \frac{\rho U_0^2 S_3}{2 \mathcal{M}_{xz}^{(5)}} \int_0^L \left(\frac{\partial C_L(\ell)}{\partial p} \right)_0^e \varphi_{xz}^{(5)} d\ell \right\} s.$$

$$A_{5,20} = \left\{ -\frac{M_R}{\mathcal{M}_{xz}^{(5)}} (L - \ell_{CG}) \varphi_{xz}^{(5)}(\ell_R) \right\} s^2 + \left\{ \frac{\rho U_0^2 S_3}{2 \mathcal{M}_{xz}^{(5)}} \int_0^L \left(\frac{\partial C_L(\ell)}{\partial q} \right)_0^e \varphi_{xz}^{(5)} d\ell \right\} s.$$

$$A_{5'21} = \left\{ \frac{\rho U_0^2 S_3}{2 \mathcal{M}_{xz}^{(5)}} \int_0^L \left(\frac{\partial C_L(\ell)}{\partial r} \right)_0^e \varphi_{xz}^{(5)} d\ell \right\} s$$

$$A_{5,22} = \left\{ \frac{-M_{C_1}}{\mathcal{M}_{xz}^{(5)}} \sigma_{xz}^{(5)}(X_{1B}) \right\} s^2 + \left\{ \frac{K_1}{\mathcal{M}_{xz}^{(5)}} \varphi_{xz}^{(5)}(X_1) - \frac{M_{I_1}(T_s + T_c - D)}{M_t \mathcal{M}_{xz}^{(5)}} \sigma_{xz}^{(5)}(X_{1B}) \right\}$$

$$A_{5'23} = \left\{ \frac{-M_{C_2}}{\mathcal{M}_{xz}^{(5)}} \sigma_{xz}^{(5)}(X_{2B}) \right\} s^2 + \left\{ \frac{K_2}{\mathcal{M}_{xz}^{(5)}} \varphi_{xz}^{(5)}(X_2) - \frac{M_{I_2}(T_s + T_c - D)}{M_t \mathcal{M}_{xz}^{(5)}} \sigma_{xz}^{(5)}(X_{2B}) \right\}$$

$$A_{5'24} = \left\{ \frac{-M_{C_3}}{\mathcal{M}_{xz}^{(5)}} \sigma_{xz}^{(5)}(X_{3B}) \right\} s^2 + \left\{ \frac{K_3}{\mathcal{M}_{xz}^{(5)}} \varphi_{xz}^{(5)}(X_3) - \frac{M_{I_3}(T_s+T_c-D)}{M_t\, \mathcal{M}_{xz}^{(5)}} \sigma_{xz}^{(5)}(X_{3,B}) \right\}.$$

$$A_{5'25} = \left\{ -\frac{M_{C_4}}{\mathcal{M}_{xz}^{(5)}} \sigma_{xz}^{(5)}(X_{4B}) \right\} s^2 + \left\{ \frac{K_4}{\mathcal{M}_{xz}^{(5)}} \varphi_{xz}^{(5)}(X_4) - \frac{M_{I_4}(T_s+T_c-D)}{M_t\, \mathcal{M}_{xz}^{(5)}} \sigma_{xz}^{(5)}(X_{4,B}) \right\}$$

$$A_{5'26} \cdots\cdots A_{5'29} = 0.$$

$$A_{5'30} = \left\{ \frac{M_R\, \ell_R}{\mathcal{M}_{xz}^{(5)}} \varphi_{xz}^{(5)}(\ell_R) \right\} s^2 + \left\{ \frac{T_c}{\mathcal{M}_{xz}^{(5)}} \varphi_{xz}^{(5)}(eG) \right\}.$$

$$A_{5'31} \cdots\cdots A_{5'40} = 0.$$

$$A_{6'1} = \left\{ \frac{\rho U_0^2 S_2}{2\mathcal{M}_{xy}^{(1)}} \int_0^L \left[\left(\frac{\partial C_y(\ell)}{\partial w}\right)_0^e + \left(\frac{\partial C_D(\ell)}{\partial w}\right)_0^e \beta_0 \right] \varphi_{xz}^{(1)} \varphi_{xy}^{(1)} d\ell \right\} s$$

$$+ \left\{ \frac{-\rho U_0^3 S_2}{2\mathcal{M}_{xy}^{(1)}} \int_0^L \left[\left(\frac{\partial C_y(\ell)}{\partial w}\right)_0^e + \left(\frac{\partial C_D(\ell)}{\partial w}\right)_0^e \beta_0 \right] \sigma_{xz}^{(1)} \varphi_{xy}^{(1)} d\ell \right\}.$$

$$A_{6'2} = \left\{ \frac{\rho U_0^2 S_2}{2\mathcal{M}_{xy}^{(1)}} \int_0^L \left[\left(\frac{\partial C_y(\ell)}{\partial w}\right)_0^e + \left(\frac{\partial C_D(\ell)}{\partial w}\right)_0^e \beta_0 \right] \varphi_{xz}^{(2)} \varphi_{xy}^{(1)} d\ell \right\} s$$

$$+ \left\{ \frac{-\rho U_0^3 S_2}{2\mathcal{M}_{xy}^{(1)}} \int_0^L \left[\left(\frac{\partial C_y(\ell)}{\partial w}\right)_0^e + \left(\frac{\partial C_D(\ell)}{\partial w}\right)_0^e \beta_0 \right] \sigma_{xz}^{(2)} \varphi_{xy}^{(1)} d\ell \right\}.$$

$$A_{6'3} = \left\{ \frac{\rho U_0^2 S_2}{2\mathcal{M}_{xy}^{(1)}} \int_0^L \left[\left(\frac{\partial C_y(\ell)}{\partial w}\right)_0^e + \left(\frac{\partial C_D(\ell)}{\partial w}\right)_0^e \beta_0 \right] \varphi_{xz}^{(3)} \varphi_{xy}^{(1)} d\ell \right\} s$$

$$+ \left\{ \frac{-\rho U_0^3 S_2}{2\mathcal{M}_{xy}^{(1)}} \int_0^L \left[\left(\frac{\partial C_y(\ell)}{\partial w}\right)_0^e + \left(\frac{\partial C_D(\ell)}{\partial w}\right)_0^e \beta_0 \right] \sigma_{xz}^{(3)} \varphi_{xy}^{(1)} d\ell \right\}.$$

$$A_{6'4} = \left\{ \frac{\rho U_0^2 S_2}{2\mathcal{M}_{xy}^{(1)}} \int_0^L \left[\left(\frac{\partial C_y(\ell)}{\partial w}\right)_0^e + \left(\frac{\partial C_D(\ell)}{\partial w}\right)_0^e \beta_0 \right] \varphi_{xz}^{(4)} \varphi_{xy}^{(1)} d\ell \right\} s$$

$$+ \left\{ \frac{-\rho U_0^3 S_2}{2\mathcal{M}_{xy}^{(1)}} \int_0^L \left[\left(\frac{\partial C_y(\ell)}{\partial w}\right)_0^e + \left(\frac{\partial C_D(\ell)}{\partial w}\right)_0^e \beta_0 \right] \sigma_{xz}^{(4)} \varphi_{xy}^{(1)} d\ell \right\}.$$

$$A_{6,5} = \left\{ \frac{\rho U_0^2 S_2}{2\mathcal{M}_{xy}^{(1)}} \int_0^L \left[\left(\frac{\partial C_y(\ell)}{\partial w}\right)_0^e + \left(\frac{\partial C_D(\ell)}{\partial w}\right)_0^e \beta_0 \right] \varphi_{xz}^{(5)} \varphi_{xy}^{(1)} d\ell \right\} s$$

$$+ \left\{ \frac{-\rho U_0^3 S_2}{2\mathcal{M}_{xy}^{(1)}} \int_0^L \left[\left(\frac{\partial C_y(\ell)}{\partial w}\right)_0^e + \left(\frac{\partial C_D(\ell)}{\partial w}\right)_0^e \beta_0 \right] \sigma_{xz}^{(5)} \varphi_{xy}^{(1)} d\ell \right\}.$$

$$A_{6,6} = \left\{ 1 + \frac{M_R}{\mathcal{M}_{xy}^{(1)}} \left(\varphi_{xy}^{(1)}(eG) - \ell_R \sigma_{xy}^{(1)}(eG) \right) \varphi_{xy}^{(1)}(\ell_R) \right\} s^2 + \left\{ 2\xi_{xy}^{(1)} \omega_{xy}^{(1)} \right.$$

$$\left. + \frac{\rho U_0^2 S_2}{2\mathcal{M}_{xy}^{(1)}} \int_0^L \left[\left(\frac{\partial C_y(\ell)}{\partial v}\right)_0^e + \left(\frac{\partial C_D(\ell)}{\partial v}\right)_0^e \beta_0 + \frac{1}{U_0} C_D(\ell)_0^e \right] \varphi_{xy}^{(1)2} d\ell \right\} s + \left\{ \omega_{xy}^{(1)2} \right.$$

$$- \frac{(T_s + T_c)}{\mathcal{M}_{xy}^{(1)}} \sigma_{xy}^{(1)}(eG) \varphi_{xy}^{(1)}(eG) - \frac{\rho U_0^3 S_2}{2\mathcal{M}_{xy}^{(1)}} \int_0^L \left[\left(\frac{\partial C_y(\ell)}{\partial v}\right)_0^e + \left(\frac{\partial C_D(\ell)}{\partial v}\right)_0^e \beta_0 \right.$$

$$\left. \left. + \frac{1}{U_0} C_D(\ell)_0^e \right] \sigma_{xy}^{(1)} \varphi_{xy}^{(1)} d\ell \right\}.$$

$$A_{6,7} = \left\{ \frac{M_R}{\mathcal{M}_{xy}^{(1)}} \left(\varphi_{xy}^{(2)}(eG) - \ell_R \sigma_{xy}^{(2)}(eG) \right) \varphi_{xy}^{(1)}(\ell_R) \right\} s^2 + \left\{ \frac{\rho U_0^2 S_2}{2\mathcal{M}_{xy}^{(1)}} \int_0^L \left[\left(\frac{\partial C_y(\ell)}{\partial v}\right)_0^e + \left(\frac{\partial C_D(\ell)}{\partial v}\right)_0^e \beta_0 \right. \right.$$

$$\left. \left. + \frac{1}{U_0} C_D(\ell)_0^e \right] \varphi_{xy}^{(2)} \varphi_{xy}^{(1)} d\ell \right\} s + \left\{ - \frac{(T_s + T_c)}{\mathcal{M}_{xy}^{(1)}} \sigma_{xy}^{(2)}(eG) \varphi_{xy}^{(1)}(eG) \right.$$

$$\left. - \frac{\rho U_0^3 S_2}{2\mathcal{M}_{xy}^{(1)}} \int_0^L \left[\left(\frac{\partial C_y(\ell)}{\partial v}\right)_0^e + \left(\frac{\partial C_D(\ell)}{\partial v}\right)_0^e \beta_0 + \frac{1}{U_0} C_D(\ell)_0^e \right] \sigma_{xy}^{(2)} \varphi_{xy}^{(1)} d\ell \right\}.$$

$$A_{6,c} = \left\{ \frac{M_R}{\mathcal{M}_{xy}^{(1)}} \left(\varphi_{xy}^{(3)}(eG) - \ell_R \sigma_{xy}^{(3)}(eG) \right) \varphi_{xy}^{(1)}(\ell_R) \right\} s^2 + \left\{ \frac{\rho U_0^2 S_2}{2\mathcal{M}_{xy}^{(1)}} \int_0^L \left[\left(\frac{\partial C_y(\ell)}{\partial v}\right)_0^e + \left(\frac{\partial C_D(\ell)}{\partial v}\right)_0^e \beta_0 \right. \right.$$

$$\left. \left. + \frac{1}{U_0} C_D(\ell)_0^e \right] \varphi_{xy}^{(3)} \varphi_{xy}^{(1)} d\ell \right\} s + \left\{ - \frac{(T_s + T_c)}{\mathcal{M}_{xy}^{(1)}} \sigma_{xy}^{(3)}(eG) \varphi_{xy}^{(1)}(eG) \right.$$

$$\left. - \frac{\rho U_0^3 S_2}{2\mathcal{M}_{xy}^{(1)}} \int_0^L \left[\left(\frac{\partial C_y(\ell)}{\partial v}\right)_0^e + \left(\frac{\partial C_D(\ell)}{\partial v}\right)_0^e \beta_0 + \frac{1}{U_0} C_D(\ell)_0^e \right] \sigma_{xy}^{(3)} \varphi_{xy}^{(1)} d\ell \right\}.$$

$$A_{6,9} = \left\{ \frac{M_R}{\mathcal{M}_{xy}^{(1)}} \left(\varphi_{xy}^{(4)}(eG) - \ell_R\, \sigma_{xy}^{(4)}(eG) \right) \varphi_{xy}^{(1)}(\ell_R) \right\} s^2 + \left\{ \frac{\rho U_0^2 S_2}{2\mathcal{M}_{xy}^{(1)}} \int_0^L \left[\left(\frac{\partial C_y(\ell)}{\partial v} \right)_0^e + \left(\frac{\partial C_D(\ell)}{\partial v} \right)_0^e \beta_0 \right.\right.$$

$$\left.\left. + \frac{1}{U_0} C_D(\ell)_0^e \right] \varphi_{xy}^{(4)} \varphi_{xy}^{(1)} d\ell \right\} s + \left\{ - \frac{(T_s + T_c)}{\mathcal{M}_{xy}^{(1)}} \sigma_{xy}^{(4)}(eG)\, \varphi_{xy}^{(1)}(eG) \right.$$

$$\left. - \frac{\rho U_0^3 S_2}{2\mathcal{M}_{xy}^{(1)}} \int_0^L \left[\left(\frac{\partial C_y(\ell)}{\partial v} \right)_0^e + \left(\frac{\partial C_D(\ell)}{\partial v} \right)_0^e \beta_0 + \frac{1}{U_0} C_D(\ell)_0^e \right] \sigma_{xy}^{(4)} \varphi_{xy}^{(1)} d\ell \right\} .$$

$$A_{6,10} = \left\{ \frac{M_R}{\mathcal{M}_{xy}^{(1)}} \left(\varphi_{xy}^{(5)}(eG) - \ell_R\, \sigma_{xy}^{(5)}(eG) \right) \varphi_{xy}^{(1)}(\ell_R) \right\} s^2 + \left\{ \frac{\rho U_0^2 S_2}{2\mathcal{M}_{xy}^{(1)}} \int_0^L \left[\left(\frac{\partial C_y(\ell)}{\partial v} \right)_0^e + \left(\frac{\partial C_D(\ell)}{\partial v} \right)_0^e \beta_0 \right.\right.$$

$$\left.\left. + \frac{1}{U_0} C_D(\ell)_0^e \right] \varphi_{xy}^{(5)} \varphi_{xy}^{(1)} d\ell \right\} s + \left\{ - \frac{(T_s + T_c)}{\mathcal{M}_{xy}^{(1)}} \sigma_{xy}^{(5)}(eG)\, \varphi_{xy}^{(1)}(eG) \right.$$

$$\left. - \frac{\rho U_0^3 S_2}{2\mathcal{M}_{xy}^{(1)}} \int_0^L \left[\left(\frac{\partial C_y(\ell)}{\partial v} \right)_0^e + \left(\frac{\partial C_D(\ell)}{\partial v} \right)_0^e \beta_0 + \frac{1}{U_0} C_D(\ell)_0^e \right] \sigma_{xy}^{(5)} \varphi_{xy}^{(1)} d\ell \right\} .$$

$$A_{6,11} = \left\{ \frac{\rho U_0^2 S_2}{2\mathcal{M}_{xy}^{(1)}} \int_0^L \left[\left(\frac{\partial C_y(\ell)}{\partial p} \right)_0^e + \left(\frac{\partial C_D(\ell)}{\partial p} \right)_0^e \beta_0 \right] \varphi_{zy}^{(1)} \varphi_{xy}^{(1)} d\ell \right\} s$$

$$A_{6,12} = \left\{ \frac{\rho U_0^2 S_2}{2\mathcal{M}_{xy}^{(1)}} \int_0^L \left[\left(\frac{\partial C_y(\ell)}{\partial p} \right)_0^e + \left(\frac{\partial C_D(\ell)}{\partial p} \right)_0^e \beta_0 \right] \varphi_{zy}^{(2)} \varphi_{xy}^{(1)} d\ell \right\} s$$

$$A_{6,13} = \left\{ \frac{\rho U_0^2 S_2}{2\mathcal{M}_{xy}^{(1)}} \int_0^L \left[\left(\frac{\partial C_y(\ell)}{\partial p} \right)_0^e + \left(\frac{\partial C_D(\ell)}{\partial p} \right)_0^e \beta_0 \right] \varphi_{zy}^{(3)} \varphi_{xy}^{(1)} d\ell \right\} s$$

$$A_{6,14} = \left\{ \frac{\rho U_0^2 S_2}{2\mathcal{M}_{xy}^{(1)}} \int_0^L \left[\left(\frac{\partial C_y(\ell)}{\partial p} \right)_0^e + \left(\frac{\partial C_D(\ell)}{\partial p} \right)_0^e \beta_0 \right] \varphi_{zy}^{(4)} \varphi_{xy}^{(1)} d\ell \right\} s$$

$$A_{6,15} = \left\{ \frac{\rho U_0^2 S_2}{2\mathcal{M}_{xy}^{(1)}} \int_0^L \left[\left(\frac{\partial C_y(\ell)}{\partial p} \right)_0^e + \left(\frac{\partial C_D(\ell)}{\partial p} \right)_0^e \beta_0 \right] \varphi_{zy}^{(5)} \varphi_{xy}^{(1)} d\ell \right\} s$$

$$A_{6,16} = \left\{ \frac{\rho U_0^2 S_2}{2\mathcal{M}_{xy}^{(1)}} \int_0^L \left[\left(\frac{\partial C_y(\ell)}{\partial u}\right)_0^e + \left(\frac{\partial C_D(\ell)}{\partial u}\right)_0^e \beta_0 + \frac{2}{U_0} C_y(\ell)_0^e + \frac{2}{U_0} \beta_0 C_D(\ell)_0^e \right] \varphi_{xy}^{(1)} d\ell \right\}$$

$$A_{6,17} = \left\{ \frac{M_R}{\mathcal{M}_{xy}^{(1)}} (\varphi_{xy}^{(1)}(\ell_R) + \frac{\rho U_0^2 S_2}{2\mathcal{M}_{xy}^{(1)}} \int_0^L \left[\left(\frac{\partial C_y(\ell)}{\partial \dot{v}}\right)_0^e + \left(\frac{\partial C_D(\ell)}{\partial \dot{v}}\right)_0^e \beta_0 \right] \varphi_{xy}^{(1)} d\ell \right\} s$$

$$+ \left\{ \frac{\rho U_0^2 S_2}{2\mathcal{M}_{xy}^{(1)}} \int_0^L \left[\left(\frac{\partial C_y(\ell)}{\partial v}\right)_0^e + \left(\frac{\partial C_D(\ell)}{\partial v}\right)_0^e \beta_0 + \frac{1}{U_0} C_D(\ell)_0^e \right] \varphi_{xy}^{(1)} d\ell \right\}.$$

$$A_{6,18} = \left\{ \frac{\rho U_0^2 S_2}{2\mathcal{M}_{xy}^{(1)}} \int_0^L \left[\left(\frac{\partial C_y(\ell)}{\partial w}\right)_0^e + \left(\frac{\partial C_D(\ell)}{\partial w}\right)_0^e \beta_0 \right] \varphi_{xy}^{(1)} d\ell \right\}.$$

$$A_{6,19} = \left\{ \frac{\rho U_0^2 S_2}{2\mathcal{M}_{xy}^{(1)}} \int_0^L \left[\left(\frac{\partial C_y(\ell)}{\partial p}\right)_0^e + \left(\frac{\partial C_D(\ell)}{\partial p}\right)_0^e \beta_0 \right] \varphi_{xy}^{(1)} d\ell \right\} s.$$

$$A_{6,20} = \left\{ \frac{\rho U_0^2 S_2}{2\mathcal{M}_{xy}^{(1)}} \int_0^L \left[\left(\frac{\partial C_y(\ell)}{\partial q}\right)_0^e + \left(\frac{\partial C_D(\ell)}{\partial q}\right)_0^e \beta_0 \right] \varphi_{xy}^{(1)} d\ell \right\} s.$$

$$A_{6,21} = \left\{ \frac{-M_R}{\mathcal{M}_{xy}^{(1)}} (L - \ell_{CG}) \varphi_{xy}^{(1)}(\ell_R) \right\} s^2 + \left\{ \frac{\rho U_0^2 S_2}{2\mathcal{M}_{xy}^{(1)}} \int_0^L \left[\left(\frac{\partial C_y(\ell)}{\partial r}\right)_0^e \right.\right.$$

$$\left.\left. + \left(\frac{\partial C_D(\ell)}{\partial r}\right)_0^e \beta_0 \right] \varphi_{xy}^{(1)} d\ell \right\} s.$$

$$A_{6,22} \cdots\cdots A_{6,25} = 0.$$

$$A_{6,26} = \left\{ \frac{M_{c_1}}{\mathcal{M}_{xy}^{(1)}} \sigma_{xy}^{(1)}(X_{1B}) \right\} s^2 + \left\{ \frac{K_1}{\mathcal{M}_{xy}^{(1)}} \varphi_{xy}^{(1)}(X_1) + \frac{M_{I_1}(T_s + T_c - D)}{M_t \mathcal{M}_{xy}^{(1)}} \sigma_{xy}^{(1)}(X_{1B}) \right\}.$$

$$A_{6,27} = \left\{ \frac{M_{c_2}}{\mathcal{M}_{xy}^{(1)}} \sigma_{xy}^{(1)}(X_{2B}) \right\} s^2 + \left\{ \frac{K_2}{\mathcal{M}_{xy}^{(1)}} \varphi_{xy}^{(1)}(X_2) + \frac{M_{I_2}(T_s + T_c - D)}{M_t \mathcal{M}_{xy}^{(1)}} \sigma_{xy}^{(1)}(X_{2B}) \right\}$$

$$A_{6,28} = \left\{ \frac{M_{c_3}}{\mathcal{M}_{xy}^{(1)}} \sigma_{xy}^{(1)}(X_{3B}) \right\} s^2 + \left\{ \frac{K_3}{\mathcal{M}_{xy}^{(1)}} \varphi_{xy}^{(1)}(X_3) + \frac{M_{I_3}(T_s + T_c - D)}{M_t \mathcal{M}_{xy}^{(1)}} \sigma_{xy}^{(1)}(X_{3B}) \right\}$$

$$A_{6,29} = \left\{ \frac{M_{C_4}}{\mathcal{M}^{(1)}_{xy}} \sigma^{(1)}_{xy}(X_{4B}) \right\} s^2 + \left\{ \frac{K_4}{\mathcal{M}^{(1)}_{xy}} \varphi^{(1)}_{xy}(X_4) + \frac{M_{I_4}(T_s+T_c-D)}{M_t\,\mathcal{M}^{(1)}_{xy}} \sigma^{(1)}_{xy}(X_{4B}) \right\}$$

$$A_{6,30} = 0.$$

$$A_{6,31} = \left\{ \frac{M_R\,\ell_R}{\mathcal{M}^{(1)}_{xy}} \varphi^{(1)}_{xy}(\ell_R) \right\} s^2 + \left\{ \frac{T_c}{\mathcal{M}^{(1)}_{xy}} \varphi^{(1)}_{xy}(eG) \right\}.$$

$$A_{6,32} \cdots\cdots A_{6,40} = 0.$$

$$A_{7,1} = \left\{ \frac{\rho U_0^2 S_2}{2\mathcal{M}^{(2)}_{xy}} \int_0^L \left[\left(\frac{\partial C_y(\ell)}{\partial w}\right)^e_0 + \left(\frac{\partial C_D(\ell)}{\partial w}\right)^e_0 \beta_0 \right] \varphi^{(1)}_{xz}\,\varphi^{(2)}_{xy}\,d\ell \right\} s$$

$$+ \left\{ \frac{-\rho U_0^3 S_2}{2\mathcal{M}^{(2)}_{xy}} \int_0^L \left[\left(\frac{\partial C_y(\ell)}{\partial w}\right)^e_0 + \left(\frac{\partial C_D(\ell)}{\partial w}\right)^e_0 \beta_0 \right] \sigma^{(1)}_{xz}\,\varphi^{(2)}_{xy}\,d\ell \right\}.$$

$$A_{7,2} = \left\{ \frac{\rho U_0^2 S_2}{2\mathcal{M}^{(2)}_{xy}} \int_0^L \left[\left(\frac{\partial C_y(\ell)}{\partial w}\right)^e_0 + \left(\frac{\partial C_D(\ell)}{\partial w}\right)^e_0 \beta_0 \right] \varphi^{(2)}_{xz}\,\varphi^{(2)}_{xy}\,d\ell \right\} s$$

$$+ \left\{ \frac{-\rho U_0^3 S_2}{2\mathcal{M}^{(2)}_{xy}} \int_0^L \left[\left(\frac{\partial C_y(\ell)}{\partial w}\right)^e_0 + \left(\frac{\partial C_D(\ell)}{\partial w}\right)^e_0 \beta_0 \right] \sigma^{(2)}_{xz}\,\varphi^{(2)}_{xy}\,d\ell \right\}.$$

$$A_{7,3} = \left\{ \frac{\rho U_0^2 S_2}{2\mathcal{M}^{(2)}_{xy}} \int_0^L \left[\left(\frac{\partial C_y(\ell)}{\partial w}\right)^e_0 + \left(\frac{\partial C_D(\ell)}{\partial w}\right)^e_0 \beta_0 \right] \varphi^{(3)}_{xz}\,\varphi^{(2)}_{xy}\,d\ell \right\} s$$

$$+ \left\{ \frac{-\rho U_0^3 S_2}{2\mathcal{M}^{(2)}_{xy}} \int_0^L \left[\left(\frac{\partial C_y(\ell)}{\partial w}\right)^e_0 + \left(\frac{\partial C_D(\ell)}{\partial w}\right)^e_0 \beta_0 \right] \sigma^{(3)}_{xz}\,\varphi^{(2)}_{xy}\,d\ell \right\}.$$

$$A_{7,4} = \left\{ \frac{\rho U_0^2 S_2}{2\mathcal{M}^{(2)}_{xy}} \int_0^L \left[\left(\frac{\partial C_y(\ell)}{\partial w}\right)^e_0 + \left(\frac{\partial C_D(\ell)}{\partial w}\right)^e_0 \beta_0 \right] \varphi^{(4)}_{xz}\,\varphi^{(2)}_{xy}\,d\ell \right\} s$$

$$+ \left\{ \frac{-\rho U_0^3 S_2}{2\mathcal{M}^{(2)}_{xy}} \int_0^L \left[\left(\frac{\partial C_y(\ell)}{\partial w}\right)^e_0 + \left(\frac{\partial C_D(\ell)}{\partial w}\right)^e_0 \beta_0 \right] \sigma^{(4)}_{xz}\,\varphi^{(2)}_{xy}\,d\ell \right\}.$$

$$A_{7,5} = \left\{ \frac{\rho U_0^2 S_2}{2\mathcal{M}_{xy}^{(2)}} \int_0^L \left[\left(\frac{\partial C_y(\ell)}{\partial w} \right)_0^e + \left(\frac{\partial C_D(\ell)}{\partial w} \right)_0^e \beta_0 \right] \varphi_{xz}^{(5)} \varphi_{xy}^{(2)} d\ell \right\} s$$

$$+ \left\{ \frac{-\rho U_0^3 S_2}{2\mathcal{M}_{xy}^{(2)}} \int_0^L \left[\left(\frac{\partial C_y(\ell)}{\partial w} \right)_0^e + \left(\frac{\partial C_D(\ell)}{\partial w} \right)_0^e \beta_0 \right] \sigma_{xz}^{(5)} \varphi_{xy}^{(2)} d\ell \right\} .$$

$$A_{7,6} = \left\{ \frac{M_R}{\mathcal{M}_{xy}^{(2)}} \left(\varphi_{xy}^{(1)}(eG) - \ell_R \sigma_{xy}^{(1)}(eG) \right) \varphi_{xy}^{(2)}(\ell_R) \right\} s^2 + \left\{ \frac{\rho U_0^2 S_2}{2\mathcal{M}_{xy}^{(2)}} \int_0^L \left[\left(\frac{\partial C_y(\ell)}{\partial v} \right)_0^e + \left(\frac{\partial C_D(\ell)}{\partial v} \right)_0^e \beta_0 \right.\right.$$

$$\left.\left. + \frac{1}{U_0} C_D(\ell)_0^e \right] \varphi_{xy}^{(1)} \varphi_{xy}^{(2)} d\ell \right\} s + \left\{ - \frac{(T_s + T_c)}{\mathcal{M}_{xy}^{(2)}} \sigma_{xy}^{(1)}(eG) \varphi_{xy}^{(2)}(eG) \right.$$

$$\left. - \frac{\rho U_0^3 S_2}{2\mathcal{M}_{xy}^{(2)}} \int_0^L \left[\left(\frac{\partial C_y(\ell)}{\partial v} \right)_0^e + \left(\frac{\partial C_D(\ell)}{\partial v} \right)_0^e \beta_0 + \frac{1}{U_0} C_D(\ell)_0^e \right] \sigma_{xy}^{(1)} \varphi_{xy}^{(2)} d\ell \right\} .$$

$$A_{7,7} = \left\{ 1 + \frac{M_R}{\mathcal{M}_{xy}^{(2)}} \left(\varphi_{xy}^{(2)}(eG) - \ell_R \sigma_{xy}^{(2)}(eG) \right) \varphi_{xy}^{(2)}(\ell_R) \right\} s^2 + \left\{ 2 \xi_{xy}^{(2)} \omega_{xy}^{(2)} \right.$$

$$\left. + \frac{\rho U_0^2 S_2}{2\mathcal{M}_{xy}^{(2)}} \int_0^L \left[\left(\frac{\partial C_y(\ell)}{\partial v} \right)_0^e + \left(\frac{\partial C_D(\ell)}{\partial v} \right)_0^e \beta_0 + \frac{1}{U_0} C_D(\ell)_0^e \right] \varphi_{xy}^{(2)2} d\ell \right\} s + \left\{ \omega_{xy}^{(2)2} \right.$$

$$- \frac{(T_s + T_c)}{\mathcal{M}_{xy}^{(2)}} \sigma_{xy}^{(2)}(eG) \varphi_{xy}^{(2)}(eG) - \frac{\rho U_0^3 S_2}{2\mathcal{M}_{xy}^{(2)}} \int_0^L \left[\left(\frac{\partial C_y(\ell)}{\partial v} \right)_0^e + \left(\frac{\partial C_D(\ell)}{\partial v} \right)_0^e \beta_0 \right.$$

$$\left.\left. + \frac{1}{U_0} C_D(\ell)_0^e \right] \sigma_{xy}^{(2)} \varphi_{xy}^{(2)} d\ell \right\} .$$

$$A_{7,8} = \left\{ \frac{M_R}{\mathcal{M}_{xy}^{(2)}} \left(\varphi_{xy}^{(3)}(eG) - \ell_R \sigma_{xy}^{(3)}(eG) \right) \varphi_{xy}^{(2)}(\ell_R) \right\} s^2 + \left\{ \frac{\rho U_0^2 S_2}{2\mathcal{M}_{xy}^{(2)}} \int_0^L \left[\left(\frac{\partial C_y(\ell)}{\partial v} \right)_0^e + \left(\frac{\partial C_D(\ell)}{\partial v} \right)_0^e \beta_0 \right.\right.$$

$$\left.\left. + \frac{1}{U_0} C_D(\ell)_0^e \right] \varphi_{xy}^{(3)} \varphi_{xy}^{(2)} d\ell \right\} s + \left\{ - \frac{(T_s + T_c)}{\mathcal{M}_{xy}^{(2)}} \sigma_{xy}^{(3)}(eG) \varphi_{xy}^{(2)}(eG) \right.$$

$$\left. - \frac{\rho U_0^3 S_2}{2\mathcal{M}_{xy}^{(2)}} \int_0^L \left[\left(\frac{\partial C_y(\ell)}{\partial v} \right)_0^e + \left(\frac{\partial C_D(\ell)}{\partial v} \right)_0^e \beta_0 + \frac{1}{U_0} C_D(\ell)_0^e \right] \sigma_{xy}^{(3)} \varphi_{xy}^{(2)} d\ell \right\} .$$

$$A_{7,9} = \left\{ \frac{M_R}{\mathcal{M}^{(2)}_{xy}} \left(\varphi^{(4)}_{xy}(eG) - \ell_R\, \sigma^{(4)}_{xy}(eG) \right) \varphi^{(2)}_{xy}(\ell_R) \right\} s^2 + \left\{ \frac{\rho U_0^2 S_2}{2\mathcal{M}^{(2)}_{xy}} \int_0^L \left[\left(\frac{\partial C_y(\ell)}{\partial v} \right)^e_0 + \left(\frac{\partial C_D(\ell)}{\partial v} \right)^e_0 \beta_0 \right.\right.$$

$$\left.\left. + \frac{1}{U_0} C_D(\ell)^e_0 \right] \varphi^{(4)}_{xy} \varphi^{(2)}_{xy} d\ell \right\} s + \left\{ - \frac{(T_s + T_c)}{\mathcal{M}^{(2)}_{xy}} \sigma^{(4)}_{xy}(eG)\, \varphi^{(2)}_{xy}(eG) \right.$$

$$\left. - \frac{\rho U_0^3 S_2}{2\mathcal{M}^{(2)}_{xy}} \int_0^L \left[\left(\frac{\partial C_y(\ell)}{\partial v} \right)^e_0 + \left(\frac{\partial C_D(\ell)}{\partial v} \right)^e_0 \beta_0 + \frac{1}{U_0} C_D(\ell)^e_0 \right] \sigma^{(4)}_{xy}\, \varphi^{(2)}_{xy}\, d\ell \right\} .$$

$$A_{7,10} = \left\{ \frac{M_R}{\mathcal{M}^{(2)}_{xy}} \left(\varphi^{(5)}_{xy}(eG) - \ell_R\, \sigma^{(5)}_{xy}(eG) \right) \varphi^{(2)}_{xy}(\ell_R) \right\} s^2 + \left\{ \frac{\rho U_0^2 S_2}{2\mathcal{M}^{(2)}_{xy}} \int_0^L \left[\left(\frac{\partial C_y(\ell)}{\partial v} \right)^e_0 + \left(\frac{\partial C_D(\ell)}{\partial v} \right)^e_0 \beta_0 \right.\right.$$

$$\left.\left. + \frac{1}{U_0} C_D(\ell)^e_0 \right] \varphi^{(5)}_{xy} \varphi^{(2)}_{xy} d\ell \right\} s + \left\{ - \frac{(T_s + T_c)}{\mathcal{M}^{(2)}_{xy}} \sigma^{(5)}_{xy}(eG)\, \varphi^{(2)}_{xy}(eG) \right.$$

$$\left. - \frac{\rho U_0^3 S_2}{2\mathcal{M}^{(2)}_{xy}} \int_0^L \left[\left(\frac{\partial C\,(\ell)}{\partial v} \right)^e_0 + \left(\frac{\partial C_D(\ell)}{\partial v} \right)^e_0 \beta_0 + \frac{1}{U_0} C_D(\ell)^e_0 \right] \sigma^{(5)}_{xy}\, \varphi^{(2)}_{xy}\, d\ell \right\} .$$

$$A_{7,11} = \left\{ \frac{\rho U_0^2 S_2}{2\mathcal{M}^{(2)}_{xy}} \int_0^L \left[\left(\frac{\partial C_y(\ell)}{\partial p} \right)^e_0 + \left(\frac{\partial C_D(\ell)}{\partial p} \right)^e_0 \beta_0 \right] \varphi^{(1)}_{zy}\, \varphi^{(2)}_{xy}\, d\ell \right\} s$$

$$A_{7,12} \quad \left\{ \frac{\rho U_0^2 S_2}{2\mathcal{M}^{(2)}_{xy}} \int_0^L \left[\left(\frac{\partial C_y(\ell)}{\partial p} \right)^e_0 + \left(\frac{\partial C_D(\ell)}{\partial p} \right)^e_0 \beta_0 \right] \varphi^{(2)}_{zy}\, \varphi^{(2)}_{xy}\, d\ell \right\} s.$$

$$A_{7,13} = \left\{ \frac{\rho U_0^2 S_2}{2\mathcal{M}^{(2)}_{xy}} \int_0^L \left[\left(\frac{\partial C_y(\ell)}{\partial p} \right)^e_0 + \left(\frac{\partial C_D(\ell)}{\partial p} \right)^e_0 \beta_0 \right] \varphi^{(3)}_{zy}\, \varphi^{(2)}_{xy}\, d\ell \right\} s$$

$$A_{7,14} \quad \left\{ \frac{\rho U_0^2 S_2}{2\mathcal{M}^{(2)}_{xy}} \int_0^L \left[\left(\frac{\partial C_y(\ell)}{\partial p} \right)^e_0 + \left(\frac{\partial C_D(\ell)}{\partial p} \right)^e_0 \beta_0 \right] \varphi^{(4)}_{zy}\, \varphi^{(2)}_{xy}\, d\ell \right\} s.$$

$$A_{7,15} = \left\{ \frac{\rho U_0^2 S_2}{2\mathcal{M}^{(2)}_{xy}} \int_0^L \left[\left(\frac{\partial C_y(\ell)}{\partial p} \right)^e_0 + \left(\frac{\partial C_D(\ell)}{\partial p} \right)^e_0 \beta_0 \right] \varphi^{(5)}_{zy}\, \varphi^{(2)}_{xy}\, d\ell \right\} s$$

$$A_{7,16} = \left\{ \frac{\rho U_0^2 S_2}{2\mathcal{M}^{(2)}_{xy}} \int_0^L \left[\left(\frac{\partial C_y(\ell)}{\partial u} \right)^e_0 + \left(\frac{\partial C_D(\ell)}{\partial u} \right)^e_0 \beta_0 + \frac{2}{U_0} C_y(\ell)^e_0 + \frac{2}{U_0} \beta_0 C_D(\ell)^e_0 \right] \varphi^{(2)}_{xy}\, d\ell \right\} .$$

$$A_{7,17} = \left\{ \frac{M_R}{\mathcal{M}_{xy}^{(2)}} \varphi_{xy}^{(2)}(\ell_R) + \frac{\rho U_0^2 S_2}{2\mathcal{M}_{xy}^{(2)}} \int_0^L \left[\left(\frac{\partial C_y(\ell)}{\partial \dot{v}} \right)_0^e + \left(\frac{\partial C_D(\ell)}{\partial \dot{v}} \right)_0^e \beta_0 \right] \varphi_{xy}^{(2)} \, d\ell \right\} s$$

$$+ \left\{ \frac{\rho U_0^2 S_2}{2\mathcal{M}_{xy}^{(2)}} \int_0^L \left[\left(\frac{\partial C_y(\ell)}{\partial v} \right)_0^e + \left(\frac{\partial C_D(\ell)}{\partial v} \right)_0^e \beta_0 + \frac{1}{U_0} C_D(\ell)_0^e \right] \varphi_{xy}^{(2)} d\ell \right\}.$$

$$A_{7,18} = \left\{ \frac{\rho U_0^2 S_2}{2\mathcal{M}_{xy}^{(2)}} \int_0^L \left[\left(\frac{\partial C_y(\ell)}{\partial w} \right)_0^e + \left(\frac{\partial C_D(\ell)}{\partial w} \right)_0^e \beta_0 \right] \varphi_{xy}^{(2)} \, d\ell \right\}.$$

$$A_{7,19} \quad \left\{ \frac{\rho U_0^2 S_2}{2\mathcal{M}_{xy}^{(2)}} \int_0^L \left[\left(\frac{\partial C_y(\ell)}{\partial p} \right)_0^e + \left(\frac{\partial C_D(\ell)}{\partial p} \right)_0^e \beta_0 \right] \varphi_{xy}^{(2)} \, d\ell \right\} s.$$

$$A_{7,20} = \left\{ \frac{\rho U_0^2 S_2}{2\mathcal{M}_{xy}^{(2)}} \int_0^L \left[\left(\frac{\partial C_y(\ell)}{\partial q} \right)_0^e + \left(\frac{\partial C_D(\ell)}{\partial q} \right)_0^e \beta_0 \right] \varphi_{xy}^{(2)} \, d\ell \right\} s.$$

$$A_{7,21} = \left\{ \frac{-M_R}{\mathcal{M}_{xy}^{(2)}} (L - \ell_{CG}) \varphi_{xy}^{(2)}(\ell_R) \right\} s^2 + \left\{ \frac{\rho U_0^2 S}{2\mathcal{M}_{xy}^{(2)}} \int_0^L \left(\frac{\partial C_y(\ell)}{\partial r} \right)_0^e + \left(\frac{\partial C_D(\ell)}{\partial r} \right)_0^e \beta_0 \right] \varphi_{xy}^{(2)} d\ell \right\} s$$

$$A_{7,22} \quad \cdots\cdot \quad A_{7,25} = 0.$$

$$A_{7,26} = \left\{ \frac{M_{C_1}}{\mathcal{M}_{xy}^{(2)}} \sigma_{xy}^{(2)}(X_{1B}) \right\} s^2 + \left\{ \frac{K_1}{\mathcal{M}_{xy}^{(2)}} \varphi_{xy}^{(2)}(X_1) + \frac{M_{l_1}(T_s + T_c - D)}{M_t \mathcal{M}_{xy}^{(2)}} \sigma_{xy}^{(2)}(X_{1B}) \right\}.$$

$$A_{7,27} = \left\{ \frac{M_{C_2}}{\mathcal{M}_{xy}^{(2)}} \sigma_{xy}^{(2)}(X_{2B}) \right\} s^2 + \left\{ \frac{K_2}{\mathcal{M}_{xy}^{(2)}} \varphi_{xy}^{(2)}(X_2) + \frac{M_{l_2}(T_s + T_c - D)}{M_t \mathcal{M}_{xy}^{(2)}} \sigma_{xy}^{(2)}(X_{2B}) \right\}$$

$$A_{7,28} = \left\{ \frac{M_{C_3}}{\mathcal{M}_{xy}^{(2)}} \sigma_{xy}^{(2)}(X_{3B}) \right\} s^2 + \left\{ \frac{K_3}{\mathcal{M}_{xy}^{(2)}} \varphi_{xy}^{(2)}(X_3) + \frac{M_{l_3}(T_s + T_c - D)}{M_t \mathcal{M}_{xy}^{(2)}} \sigma_{xy}^{(2)}(X_{3B}) \right\}$$

$$A_{7,29} = \left\{ \frac{M_{C_4}}{\mathcal{M}_{xy}^{(2)}} \sigma_{xy}^{(2)}(X_{4B}) \right\} s^2 + \left\{ \frac{K_4}{\mathcal{M}_{xy}^{(2)}} \varphi_{xy}^{(2)}(X_4) + \frac{M_{l_4}(T_s + T_c - D)}{M_t \mathcal{M}_{xy}^{(2)}} \sigma_{xy}^{(2)}(X_{4B}) \right\}$$

$$A_{7,30} = 0.$$

$$A_{7,31} = \left\{ \frac{M_R \, \ell_R}{\mathcal{M}_{xy}^{(2)}} \varphi_{xy}^{(2)}(\ell_R) \right\} s^2 + \left\{ \frac{T_c}{\mathcal{M}_{xy}^{(2)}} \varphi_{xy}^{(2)}(eG) \right\}.$$

$$A_{7,32} \quad \cdots\cdots \quad A_{7,40} = 0.$$

$$A_{8,1} = \left\{ \frac{\rho U_0^2 S_2}{2\mathcal{M}_{xy}^{(3)}} \int_0^L \left[\left(\frac{\partial C_y(\ell)}{\partial w} \right)_0^e + \left(\frac{\partial C_D(\ell)}{\partial w} \right)_0^e \beta_0 \right] \varphi_{xz}^{(1)} \varphi_{xy}^{(3)} d\ell \right\} s$$

$$+ \left\{ \frac{-\rho U_0^3 S_2}{2\mathcal{M}_{xy}^{(3)}} \int_0^L \left[\left(\frac{\partial C_y(\ell)}{\partial w} \right)_0^e + \left(\frac{\partial C_D(\ell)}{\partial w} \right)_0^e \beta_0 \right] \sigma_{xz}^{(1)} \varphi_{xy}^{(3)} d\ell \right\}.$$

$$A_{8,2} = \left\{ \frac{\rho U_0^2 S_2}{2\mathcal{M}_{xy}^{(3)}} \int_0^L \left[\left(\frac{\partial C_y(\ell)}{\partial w} \right)_0^e + \left(\frac{\partial C_D(\ell)}{\partial w} \right)_0^e \beta_0 \right] \varphi_{xz}^{(2)} \varphi_{xy}^{(3)} d\ell \right\} s$$

$$+ \left\{ \frac{-\rho U_0^3 S_2}{2\mathcal{M}_{xy}^{(3)}} \int_0^L \left[\left(\frac{\partial C_y(\ell)}{\partial w} \right)_0^e + \left(\frac{\partial C_D(\ell)}{\partial w} \right)_0^e \beta_0 \right] \sigma_{xz}^{(2)} \varphi_{xy}^{(3)} d\ell \right\}.$$

$$A_{8,3} = \left\{ \frac{\rho U_0^2 S_2}{2\mathcal{M}_{xy}^{(3)}} \int_0^L \left[\left(\frac{\partial C_y(\ell)}{\partial w} \right)_0^e + \left(\frac{\partial C_D(\ell)}{\partial w} \right)_0^e \beta_0 \right] \varphi_{xz}^{(3)} \varphi_{xy}^{(3)} d\ell \right\} s$$

$$+ \left\{ \frac{-\rho U_0^3 S_2}{2\mathcal{M}_{xy}^{(3)}} \int_0^L \left[\left(\frac{\partial C_y(\ell)}{\partial w} \right)_0^e + \left(\frac{\partial C_D(\ell)}{\partial w} \right)_0^e \beta_0 \right] \sigma_{xz}^{(3)} \varphi_{xy}^{(3)} d\ell \right\}.$$

$$A_{8,4} = \left\{ \frac{\rho U_0^2 S_2}{2\mathcal{M}_{xy}^{(3)}} \int_0^L \left[\left(\frac{\partial C_y(\ell)}{\partial w} \right)_0^e + \left(\frac{\partial C_D(\ell)}{\partial w} \right)_0^e \beta_0 \right] \varphi_{xz}^{(4)} \varphi_{xy}^{(3)} d\ell \right\} s$$

$$+ \left\{ \frac{-\rho U_0^3 S_2}{2\mathcal{M}_{xy}^{(3)}} \int_0^L \left[\left(\frac{\partial C_y(\ell)}{\partial w} \right)_0^e + \left(\frac{\partial C_D(\ell)}{\partial w} \right)_0^e \beta_0 \right] \sigma_{xz}^{(4)} \varphi_{xy}^{(3)} d\ell \right\}.$$

$$A_{8,5} = \left\{ \frac{\rho U_0^2 S_2}{2\mathcal{M}_{xy}^{(3)}} \int_0^L \left[\left(\frac{\partial C_y(\ell)}{\partial w} \right)_0^e + \left(\frac{\partial C_D(\ell)}{\partial w} \right)_0^e \beta_0 \right] \varphi_{xz}^{(5)} \varphi_{xy}^{(3)} d\ell \right\} s$$

$$+ \left\{ \frac{-\rho U_0^3 S_2}{2\mathcal{M}_{xy}^{(3)}} \int_0^L \left[\left(\frac{\partial C_y(\ell)}{\partial w} \right)_0^e + \left(\frac{\partial C_D(\ell)}{\partial w} \right)_0^e \beta_0 \right] \sigma_{xz}^{(5)} \varphi_{xy}^{(3)} d\ell \right\}.$$

$$A_{8,6} = \left\{ \frac{M_R}{\mathcal{M}_{xy}^{(3)}} \left(\varphi_{xy}^{(1)}(eG) - \ell_R \sigma_{xy}^{(1)}(eG) \right) \varphi_{xy}^{(3)}(\ell_R) \right\} s^2 + \left\{ \frac{\rho U_0^2 S_2}{2\mathcal{M}_{xy}^{(3)}} \int_0^L \left[\left(\frac{\partial c_y(\ell)}{\partial v} \right)_0^e + \left(\frac{\partial c_D(\ell)}{\partial v} \right)_0^e \beta_0 \right.\right.$$

$$\left.\left. + \frac{1}{U_0} c_D(\ell)_0^e \right] \varphi_{xy}^{(1)} \varphi_{xy}^{(3)} d\ell \right\} s + \left\{ - \frac{(T_s + T_c)}{\mathcal{M}_{xy}^{(3)}} \sigma_{xy}^{(1)}(eG)\, \varphi_{xy}^{(3)}(eG) \right.$$

$$\left. - \frac{\rho U_0^3 S_2}{2\mathcal{M}_{xy}^{(3)}} \int_0^L \left[\left(\frac{\partial c_y(\ell)}{\partial v} \right)_0^e + \left(\frac{\partial c_D(\ell)}{\partial v} \right)_0^e \beta_0 + \frac{1}{U_0} c_D(\ell)_0^e \right] \sigma_{xy}^{(1)} \varphi_{xy}^{(3)} d\ell \right\}.$$

$$A_{8,7} = \left\{ \frac{M_R}{\mathcal{M}_{xy}^{(3)}} \left(\varphi_{xy}^{(2)}(eG) - \ell_R \sigma_{xy}^{(2)}(eG) \right) \varphi_{xy}^{(3)}(\ell_R) \right\} s^2 + \left\{ \frac{\rho U_0^2 S_2}{2\mathcal{M}_{xy}^{(3)}} \int_0^L \left[\left(\frac{\partial c_y(\ell)}{\partial v} \right)_0^e + \left(\frac{\partial c_D(\ell)}{\partial v} \right)_0^e \beta_0 \right.\right.$$

$$\left.\left. + \frac{1}{U_0} c_D(\ell)_0^e \right] \varphi_{xy}^{(2)} \varphi_{xy}^{(3)} d\ell \right\} s + \left\{ - \frac{(T_s + T_c)}{\mathcal{M}_{xy}^{(3)}} \sigma_{xy}^{(2)}(eG)\, \varphi_{xy}^{(3)}(eG) \right.$$

$$\left. - \frac{\rho U_0^3 S_2}{2\mathcal{M}_{xy}^{(3)}} \int_0^L \left[\left(\frac{\partial c_y(\ell)}{\partial v} \right)_0^e + \left(\frac{\partial c_D(\ell)}{\partial v} \right)_0^e \beta_0 + \frac{1}{U_0} c_D(\ell)_0^e \right] \sigma_{xy}^{(2)} \varphi_{xy}^{(3)} d\ell \right\}.$$

$$A_{8,8} = \left\{ 1 + \frac{M_R}{\mathcal{M}_{xy}^{(3)}} \left(\varphi_{xy}^{(3)}(eG) - \ell_R \sigma_{xy}^{(3)}(eG) \right) \varphi_{xy}^{(3)}(\ell_R) \right\} s^2 + \left\{ 2 \xi_{xy}^{(3)} \omega_{xy}^{(3)} \right.$$

$$\left. + \frac{\rho U_0^2 S_2}{2\mathcal{M}_{xy}^{(3)}} \int_0^L \left[\left(\frac{\partial c_y(\ell)}{\partial v} \right)_0^e + \left(\frac{\partial c_D(\ell)}{\partial v} \right)_0^e \beta_0 + \frac{1}{U_0} c_D(\ell)_0^e \right] \varphi_{xy}^{(3)2} d\ell \right\} s + \left\{ \omega_{xy}^{(3)2} \right.$$

$$- \frac{(T_s + T_c)}{\mathcal{M}_{xy}^{(3)}} \sigma_{xy}^{(3)}(eG)\, \varphi_{xy}^{(3)}(eG) - \frac{\rho U_0^3 S_2}{2\mathcal{M}_{xy}^{(3)}} \int_0^L \left[\left(\frac{\partial c_y(\ell)}{\partial v} \right)_0^e + \left(\frac{\partial c_D(\ell)}{\partial v} \right)_0^e \beta_0 \right.$$

$$\left.\left. + \frac{1}{U_0} c_D(\ell)_0^e \right] \sigma_{xy}^{(3)} \varphi_{xy}^{(3)} d\ell \right\}.$$

$$A_{8,9} = \left\{ \frac{M_R}{\mathcal{M}_{xy}^{(3)}} \left(\varphi_{xy}^{(4)}(eG) - \ell_R \sigma_{xy}^{(4)}(eG) \right) \varphi_{xy}^{(3)}(\ell_R) \right\} s^2 + \left\{ \frac{\rho U_0^2 S_2}{2\mathcal{M}_{xy}^{(3)}} \int_0^L \left[\left(\frac{\partial c_y(\ell)}{\partial v} \right)_0^e + \left(\frac{\partial c_D(\ell)}{\partial v} \right)_0^e \beta_0 \right.\right.$$

$$\left.\left. + \frac{1}{U_0} c_D(\ell)_0^e \right] \varphi_{xy}^{(4)} \varphi_{xy}^{(3)} d\ell \right\} s + \left\{ - \frac{(T_s + T_c)}{\mathcal{M}_{xy}^{(3)}} \sigma_{xy}^{(4)}(eG)\, \varphi_{xy}^{(3)}(eG) \right.$$

$$\left. - \frac{\rho U_0^3 S_2}{2\mathcal{M}_{xy}^{(3)}} \int_0^L \left[\left(\frac{\partial c_y(\ell)}{\partial v} \right)_0^e + \left(\frac{\partial c_D(\ell)}{\partial v} \right)_0^e \beta_0 + \frac{1}{U_0} c_D(\ell)_0^e \right] \sigma_{xy}^{(4)} \varphi_{xy}^{(3)} d\ell \right\}.$$

$$A_{8,10} = \left\{ \frac{M_R}{\mathcal{M}_{xy}^{(3)}} \left(\varphi_{xy}^{(5)}(eG) - \ell_R\, \sigma_{xy}^{(5)}(eG) \right) \varphi_{xy}^{(3)}(\ell_R) \right\} s^2 + \left\{ \frac{\rho U_0^2 S_2}{2\mathcal{M}_x^{(3)}} \int_0^L \left[\left(\frac{\partial C_y(\ell)}{\partial v} \right)_0^e + \left(\frac{\partial C_D(\ell)}{\partial v} \right)_0^e \beta_0 \right.\right.$$

$$\left.\left. + \frac{1}{U_0} C_D(\ell)_0^e \right] \varphi_{xy}^{(5)} \varphi_{xy}^{(3)} d\ell \right\} s + \left\{ - \frac{(T_s + T_c)}{\mathcal{M}_{xy}^{(3)}} \sigma_{xy}^{(5)}(eG)\, \varphi_{xy}^{(3)}(eG) \right.$$

$$\left. - \frac{\rho U_0^3 S_2}{2\mathcal{M}_{xy}^{(3)}} \int_0^L \left[\left(\frac{\partial C_y(\ell)}{\partial v} \right)_0^e + \left(\frac{\partial C_D(\ell)}{\partial v} \right)_0^e \beta_0 + \frac{1}{U_0} C_D(\ell)_0^e \right] \sigma_{xy}^{(5)} \varphi_{xy}^{(3)} d\ell \right\}.$$

$$A_{8,11} = \left\{ \frac{\rho U_0^2 S_2}{2\mathcal{M}_{xy}^{(3)}} \int_0^L \left[\left(\frac{\partial C_y(\ell)}{\partial p} \right)_0^e + \left(\frac{\partial C_D(\ell)}{\partial p} \right)_0^e \beta_0 \right] \varphi_{zy}^{(1)} \varphi_{xy}^{(3)} d\ell \right\} s$$

$$A_{8,12} = \left\{ \frac{\rho U_0^2 S_2}{2\mathcal{M}_{xy}^{(3)}} \int_0^L \left[\left(\frac{\partial C_y(\ell)}{\partial p} \right)_0^e + \left(\frac{\partial C_D(\ell)}{\partial p} \right)_0^e \beta_0 \right] \varphi_{zy}^{(2)} \varphi_{xy}^{(3)} d\ell \right\} s$$

$$A_{8,13} = \left\{ \frac{\rho U_0^2 S_2}{2\mathcal{M}_{xy}^{(3)}} \int_0^L \left[\left(\frac{\partial C_y(\ell)}{\partial p} \right)_0^e + \left(\frac{\partial C_D(\ell)}{\partial p} \right)_0^e \beta_0 \right] \varphi_{zy}^{(3)} \varphi_{xy}^{(3)} d\ell \right\} s$$

$$A_{8,14} = \left\{ \frac{\rho U_0^2 S_2}{2\mathcal{M}_{xy}^{(3)}} \int_0^L \left[\left(\frac{\partial C_y(\ell)}{\partial p} \right)_0^e + \left(\frac{\partial C_D(\ell)}{\partial p} \right)_0^e \beta_0 \right] \varphi_{zy}^{(4)} \varphi_{xy}^{(3)} d\ell \right\} s$$

$$A_{8,15} = \left\{ \frac{\rho U_0^2 S_2}{2\mathcal{M}_{xy}^{(3)}} \int_0^L \left[\left(\frac{\partial C_y(\ell)}{\partial p} \right)_0^e + \left(\frac{\partial C_D(\ell)}{\partial p} \right)_0^e \beta_0 \right] \varphi_{zy}^{(5)} \varphi_{xy}^{(3)} d\ell \right\} s$$

$$A_{8,16} = \left\{ \frac{\rho U_0^2 S_2}{2\mathcal{M}_{xy}^{(3)}} \int_0^L \left[\left(\frac{\partial C_y(\ell)}{\partial u} \right)_0^e + \left(\frac{\partial C_D(\ell)}{\partial u} \right)_0^e \beta_0 + \frac{2}{U_0} C_y(\ell)_0^e + \frac{2}{U_0} \beta_0 C_D(\ell)_0^e \right] \varphi_{xy}^{(3)} d\ell \right\}.$$

$$A_{8,17} = \left\{ \frac{M_R}{\mathcal{M}_{xy}^{(3)}} \varphi_{xy}^{(3)}(\ell_R) + \frac{\rho U_0^2 S_2}{2\mathcal{M}_{xy}^{(3)}} \int_0^L \left[\left(\frac{\partial C_y(\ell)}{\partial \dot{v}} \right)_0^e + \left(\frac{\partial C_D(\ell)}{\partial \dot{v}} \right)_0^e \beta_0 \right] \varphi_{xy}^{(3)} d\ell \right\} s$$

$$+ \left\{ \frac{\rho U_0^2 S_2}{2\mathcal{M}_{xy}^{(3)}} \int_0^L \left[\left(\frac{\partial C_y(\ell)}{\partial v} \right)_0^e + \left(\frac{\partial C_D(\ell)}{\partial v} \right)_0^e \beta_0 + \frac{1}{U_0} C_D(\ell)_0^e \right] \varphi_{xy}^{(3)} d\ell \right\}.$$

$$A_{8,18} = \left\{ \frac{\rho U_0^2 S_2}{2\mathcal{M}_{xy}^{(3)}} \int_0^L \left[\left(\frac{\partial C_y(\ell)}{\partial w} \right)_0^e + \left(\frac{\partial C_D(\ell)}{\partial w} \right)_0^e \beta_0 \right] \varphi_{xy}^{(3)} d\ell \right\}.$$

$$A_{8,19} = \left\{ \frac{\rho U_0^2 S_2}{2\mathcal{M}_{xy}^{(3)}} \int_0^L \left[\left(\frac{\partial C_y(\ell)}{\partial p}\right)_0^e + \left(\frac{\partial C_D(\ell)}{\partial p}\right)_0^e \beta_0 \right] \varphi_{xy}^{(3)} \, d\ell \right\} s$$

$$A_{8,20} \quad \left\{ \frac{\rho U_0^2 S_2}{2\mathcal{M}_{xy}^{(3)}} \int_0^L \left[\left(\frac{\partial C_y(\ell)}{\partial q}\right)_0^e + \left(\frac{\partial C_D(\ell)}{\partial q}\right)_0^e \beta_0 \right] \varphi_{xy}^{(3)} \, d\ell \right\} s.$$

$$A_{8,21} = \left\{ \frac{-M_R}{\mathcal{M}_{xy}^{(3)}} (L - \ell_{CG}) \varphi_{xy}^{(3)}(\ell_R) \right\} s^2 + \left\{ \frac{\rho U_0^2 S_2}{2\mathcal{M}_{xy}^{(3)}} \int_0^L \left(\frac{\partial C_y(\ell)}{\partial r}\right)_0^e + \left(\frac{\partial C_D(\ell)}{\partial r}\right)_0^e \beta_0 \Big] \varphi_{xy}^{(3)} \, d\ell \right\} s.$$

$$A_{8,22} \cdots A_{8,25} = 0.$$

$$A_{8,26} = \left\{ \frac{M_{C_1}}{\mathcal{M}_{xy}^{(3)}} \sigma_{xy}^{(3)}(X_{1B}) \right\} s^2 + \left\{ \frac{K_1}{\mathcal{M}_{xy}^{(3)}} \varphi_{xy}^{(3)}(X_1) + \frac{M_{I_1}(T_s + T_c - D)}{M_t \mathcal{M}_{xy}^{(3)}} \sigma_{xy}^{(3)}(X_{1B}) \right\}$$

$$A_{8,27} = \left\{ \frac{M_{C_2}}{\mathcal{M}_{xy}^{(3)}} \sigma_{xy}^{(3)}(X_{2B}) \right\} s^2 + \left\{ \frac{K_2}{\mathcal{M}_{xy}^{(3)}} \varphi_{xy}^{(3)}(X_2) + \frac{M_{I_2}(T_s + T_c - D)}{M_t \mathcal{M}_{xy}^{(3)}} \sigma_{xy}^{(3)}(X_{2B}) \right\}$$

$$A_{8,28} = \left\{ \frac{M_{C_3}}{\mathcal{M}_{xy}^{(3)}} \sigma_{xy}^{(3)}(X_{3B}) \right\} s^2 + \left\{ \frac{K_3}{\mathcal{M}_{xy}^{(3)}} \varphi_{xy}^{(3)}(X_3) + \frac{M_{I_3}(T_s + T_c - D)}{M_t \mathcal{M}_{xy}^{(3)}} \sigma_{xy}^{(3)}(X_{3B}) \right\}$$

$$A_{8,29} = \left\{ \frac{M_{C_4}}{\mathcal{M}_{xy}^{(3)}} \sigma_{xy}^{(3)}(X_{4B}) \right\} s^2 + \left\{ \frac{K_4}{\mathcal{M}_{xy}^{(3)}} \varphi_{xy}^{(3)}(X_4) + \frac{M_{I_4}(T_s + T_c - D)}{M_t \mathcal{M}_{xy}^{(3)}} \sigma_{xy}^{(3)}(X_{4B}) \right\}$$

$$A_{8,30} = 0.$$

$$A_{8,31} = \left\{ \frac{M_R \ell_R}{\mathcal{M}_{xy}^{(3)}} \varphi_{xy}^{(3)}(\ell_R) \right\} s^2 + \left\{ \frac{T_c}{\mathcal{M}_{xy}^{(3)}} \varphi_{xy}^{(3)}(eG) \right\}.$$

$$A_{8,32} \cdots A_{8,40} = 0.$$

$$A_{9,1} = \left\{ \frac{\rho U_0^2 S_2}{2\mathcal{M}_{xy}^{(4)}} \int_0^L \left[\left(\frac{\partial C_y(\ell)}{\partial w}\right)_0^e + \left(\frac{\partial C_D(\ell)}{\partial w}\right)_0^e \beta_0 \right] \varphi_{xz}^{(1)} \varphi_{xy}^{(4)} \, d\ell \right\} s$$

$$+ \left\{ \frac{-\rho U_0^3 S_2}{2\mathcal{M}_{xy}^{(4)}} \int_0^L \left[\left(\frac{\partial C_y(\ell)}{\partial w}\right)_0^e + \left(\frac{\partial C_D(\ell)}{\partial w}\right)_0^e \beta_0 \right] \sigma_{xz}^{(1)} \varphi_{xy}^{(4)} \, d\ell \right\}.$$

$$A_{9,2} = \left\{ \frac{\rho U_0^2 S_2}{2\mathcal{M}_{xy}^{(4)}} \int_0^L \left[\left(\frac{\partial C_y(\ell)}{\partial w} \right)_0^e + \left(\frac{\partial C_D(\ell)}{\partial w} \right)_0^e \beta_0 \right] \varphi_{xz}^{(2)} \varphi_{xy}^{(4)} d\ell \right\} s$$

$$+ \left\{ \frac{-\rho U_0^3 S_2}{2\mathcal{M}_{xy}^{(4)}} \int_0^L \left[\left(\frac{\partial C_y(\ell)}{\partial w} \right)_0^e + \left(\frac{\partial C_D(\ell)}{\partial w} \right)_0^e \beta_0 \right] \sigma_{xz}^{(2)} \varphi_{xy}^{(4)} d\ell \right\} .$$

$$A_{9,3} = \left\{ \frac{\rho U_0^2 S_2}{2\mathcal{M}_{xy}^{(4)}} \int_0^L \left[\left(\frac{\partial C_y(\ell)}{\partial w} \right)_0^e + \left(\frac{\partial C_D(\ell)}{\partial w} \right)_0^e \beta_0 \right] \varphi_{xz}^{(3)} \varphi_{xy}^{(4)} d\ell \right\} s$$

$$+ \left\{ \frac{-\rho U_0^3 S_2}{2\mathcal{M}_{xy}^{(4)}} \int_0^L \left[\left(\frac{\partial C_y(\ell)}{\partial w} \right)_0^e + \left(\frac{\partial C_D(\ell)}{\partial w} \right)_0^e \beta_0 \right] \sigma_{xz}^{(3)} \varphi_{xy}^{(4)} d\ell \right\} .$$

$$A_{9,4} = \left\{ \frac{\rho U_0^2 S_2}{2\mathcal{M}_{xy}^{(4)}} \int_0^L \left[\left(\frac{\partial C_y(\ell)}{\partial w} \right)_0^e + \left(\frac{\partial C_D(\ell)}{\partial w} \right)_0^e \beta_0 \right] \varphi_{xz}^{(4)} \varphi_{xy}^{(4)} d\ell \right\} s$$

$$+ \left\{ \frac{-\rho U_0^3 S_2}{2\mathcal{M}_{xy}^{(4)}} \int_0^L \left[\left(\frac{\partial C_y(\ell)}{\partial w} \right)_0^e + \left(\frac{\partial C_D(\ell)}{\partial w} \right)_0^e \beta_0 \right] \sigma_{xz}^{(4)} \varphi_{xy}^{(4)} d\ell \right\} .$$

$$A_{9,5} = \left\{ \frac{\rho U_0^2 S_2}{2\mathcal{M}_{xy}^{(4)}} \int_0^L \left[\left(\frac{\partial C_y(\ell)}{\partial w} \right)_0^e + \left(\frac{\partial C_D(\ell)}{\partial w} \right)_0^e \beta_0 \right] \varphi_{xz}^{(5)} \varphi_{xy}^{(4)} d\ell \right\} s$$

$$+ \left\{ \frac{-\rho U_0^3 S_2}{2\mathcal{M}_{xy}^{(4)}} \int_0^L \left[\left(\frac{\partial C_y(\ell)}{\partial w} \right)_0^e + \left(\frac{\partial C_D(\ell)}{\partial w} \right)_0^e \beta_0 \right] \sigma_{xz}^{(5)} \varphi_{xy}^{(4)} d\ell \right\} .$$

$$A_{9,6} = \left\{ \frac{M_R}{\mathcal{M}_{xy}^{(4)}} \left(\varphi_{xy}^{(1)}(eG) - \ell_R \sigma_{xy}^{(1)}(eG) \right) \varphi_{xy}^{(4)}(\ell_R) \right\} s^2 + \left\{ \frac{\rho U_0^2 S_2}{2\mathcal{M}_{xy}^{(4)}} \int_0^L \left[\left(\frac{\partial C_y(\ell)}{\partial v} \right)_0^e + \left(\frac{\partial C_D(\ell)}{\partial v} \right)_0^e \beta_0 \right.\right.$$

$$\left.\left. + \frac{1}{U_0} C_D(\ell)_0^e \right] \varphi_{xy}^{(1)} \varphi_{xy}^{(4)} d\ell \right\} s + \left\{ - \frac{(T_s + T_c)}{\mathcal{M}_{xy}^{(4)}} \sigma_{xy}^{(1)}(eG) \varphi_{xy}^{(4)}(eG) \right.$$

$$\left. - \frac{\rho U_0^3 S_2}{2\mathcal{M}_{xy}^{(4)}} \int_0^L \left[\left(\frac{\partial C_y(\ell)}{\partial v} \right)_0^e + \left(\frac{\partial C_D(\ell)}{\partial v} \right)_0^e \beta_0 + \frac{1}{U_0} C_D(\ell)_0^e \right] \sigma_{xy}^{(1)} \varphi_{xy}^{(4)} d\ell \right\} .$$

$$A_{9,7} = \left\{ \frac{M_R}{\mathcal{M}_{xy}^{(4)}} \left(\varphi_{xy}^{(2)}(eG) - \ell_R \sigma_{xy}^{(2)}(eG) \right) \varphi_{xy}^{(4)}(\ell_R) \right\} s^2 + \left\{ \frac{\rho U_0^2 S_2}{2\mathcal{M}_{xy}^{(4)}} \int_0^L \left[\left(\frac{\partial C_y(\ell)}{\partial v} \right)_0^e + \left(\frac{\partial C_D(\ell)}{\partial v} \right)_0^e \beta_0 \right.\right.$$

$$\left.\left. + \frac{1}{U_0} C_D(\ell)_0^e \right] \varphi_{xy}^{(2)} \varphi_{xy}^{(4)} d\ell \right\} s + \left\{ - \frac{(T_s + T_c)}{\mathcal{M}_{xy}^{(4)}} \sigma_{xy}^{(2)}(eG)\, \varphi_{xy}^{(4)}(eG) \right.$$

$$\left. - \frac{\rho U_0^3 S_2}{2\mathcal{M}_{xy}^{(4)}} \int_0^L \left[\left(\frac{\partial C_y(\ell)}{\partial v} \right)_0^e + \left(\frac{\partial C_D(\ell)}{\partial v} \right)_0^e \beta_0 + \frac{1}{U_0} C_D(\ell)_0^e \right] \sigma_{xy}^{(2)} \varphi_{xy}^{(4)} d\ell \right\}.$$

$$A_{9,8} = \left\{ \frac{M_R}{\mathcal{M}_{xy}^{(4)}} \left(\varphi_{xy}^{(3)}(eG) - \ell_R \sigma_{xy}^{(3)}(eG) \right) \varphi_{xy}^{(4)}(\ell_R) \right\} s^2 + \left\{ \frac{\rho U_0^2 S_2}{2\mathcal{M}_{xy}^{(4)}} \int_0^L \left[\left(\frac{\partial C_y(\ell)}{\partial v} \right)_0^e + \left(\frac{\partial C_D(\ell)}{\partial v} \right)_0^e \beta_0 \right.\right.$$

$$\left.\left. + \frac{1}{U_0} C_D(\ell)_0^e \right] \varphi_{xy}^{(3)} \varphi_{xy}^{(4)} d\ell \right\} s + \left\{ - \frac{(T_s + T_c)}{\mathcal{M}_{xy}^{(4)}} \sigma_{xy}^{(3)}(eG)\, \varphi_{xy}^{(4)}(eG) \right.$$

$$\left. - \frac{\rho U_0^3 S_2}{2\mathcal{M}_{xy}^{(4)}} \int_0^L \left[\left(\frac{\partial C_y(\ell)}{\partial v} \right)_0^e + \left(\frac{\partial C_D(\ell)}{\partial v} \right)_0^e \beta_0 + \frac{1}{U_0} C_D(\ell)_0^e \right] \sigma_{xy}^{(3)} \varphi_{xy}^{(4)} d\ell \right\}.$$

$$A_{9,9} = \left\{ 1 + \frac{M_R}{\mathcal{M}_{xy}^{(4)}} \left(\varphi_{xy}^{(4)}(eG) - \ell_R \sigma_{xy}^{(4)}(eG) \right) \varphi_{xy}^{(4)}(\ell_R) \right\} s^2 + \left\{ 2\xi_{xy}^{(4)} \omega_{xy}^{(4)} \right.$$

$$\left. + \frac{\rho U_0^2 S_2}{2\mathcal{M}_{xy}^{(4)}} \int_0^L \left[\left(\frac{\partial C_y(\ell)}{\partial v} \right)_0^e + \left(\frac{\partial C_D(\ell)}{\partial v} \right)_0^e \beta_0 + \frac{1}{U_0} C_D(\ell)_0^e \right] \varphi_{xy}^{(4)2} d\ell \right\} s + \left\{ \omega_{xy}^{(4)2} \right.$$

$$- \frac{(T_s + T_c)}{\mathcal{M}_{xy}^{(4)}} \sigma_{xy}^{(4)}(eG)\, \varphi_{xy}^{(4)}(eG) - \frac{\rho U_0^3 S_2}{2\mathcal{M}_{xy}^{(4)}} \int_0^L \left[\left(\frac{\partial C_y(\ell)}{\partial v} \right)_0^e + \left(\frac{\partial C_D(\ell)}{\partial v} \right)_0^e \beta_0 \right.$$

$$\left.\left. + \frac{1}{U_0} C_D(\ell)_0^e \right] \sigma_{xy}^{(4)} \varphi_{xy}^{(4)} d\ell \right\}.$$

$$A_{9,10} = \left\{ \frac{M_R}{\mathcal{M}_{xy}^{(4)}} \left(\varphi_{xy}^{(5)}(eG) - \ell_R \sigma_{xy}^{(5)}(eG) \right) \varphi_{xy}^{(4)}(\ell_R) \right\} s^2 + \left\{ \frac{\rho U_0^2 S_2}{2\mathcal{M}_{xy}^{(4)}} \int_0^L \left[\left(\frac{\partial C_y(\ell)}{\partial v} \right)_0^e + \left(\frac{\partial C_D(\ell)}{\partial v} \right)_0^e \beta_0 \right.\right.$$

$$\left.\left. + \frac{1}{U_0} C_D(\ell)_0^e \right] \varphi_{xy}^{(5)} \varphi_{xy}^{(4)} d\ell \right\} s + \left\{ - \frac{(T_s + T_c)}{\mathcal{M}_{xy}^{(4)}} \sigma_{xy}^{(5)}(eG)\, \varphi_{xy}^{(4)}(eG) \right.$$

$$\left. - \frac{\rho U_0^3 S_2}{2\mathcal{M}_{xy}^{(4)}} \int_0^L \left[\left(\frac{\partial C_y(\ell)}{\partial v} \right)_0^e + \left(\frac{\partial C_D(\ell)}{\partial v} \right)_0^e \beta_0 + \frac{1}{U_0} C_D(\ell)_0^e \right] \sigma_{xy}^{(5)} \varphi_{xy}^{(4)} d\ell \right\}.$$

$$A_{9,11} = \left\{ \frac{\rho U_0^2 S_2}{2\mathcal{M}_{xy}^{(4)}} \int_0^L \left[\left(\frac{\partial C_y(\ell)}{\partial p}\right)_0^e + \left(\frac{\partial C_D(\ell)}{\partial p}\right)_0^e \beta_0 \right] \varphi_{zy}^{(1)} \varphi_{xy}^{(4)} d\ell \right\} s$$

$$A_{9,12} = \left\{ \frac{\rho U_0^2 S_2}{2\mathcal{M}_{xy}^{(4)}} \int_0^L \left[\left(\frac{\partial C_y(\ell)}{\partial p}\right)_0^e + \left(\frac{\partial C_D(\ell)}{\partial p}\right)_0^e \beta_0 \right] \varphi_{zy}^{(2)} \varphi_{xy}^{(4)} d\ell \right\} s$$

$$A_{9,13} = \left\{ \frac{\rho U_0^2 S_2}{2\mathcal{M}_{xy}^{(4)}} \int_0^L \left[\left(\frac{\partial C_y(\ell)}{\partial p}\right)_0^e + \left(\frac{\partial C_D(\ell)}{\partial p}\right)_0^e \beta_0 \right] \varphi_{zy}^{(3)} \varphi_{xy}^{(4)} d\ell \right\} s$$

$$A_{9,14} = \left\{ \frac{\rho U_0^2 S_2}{2\mathcal{M}_{xy}^{(4)}} \int_0^L \left[\left(\frac{\partial C_y(\ell)}{\partial p}\right)_0^e + \left(\frac{\partial C_D(\ell)}{\partial p}\right)_0^e \beta_0 \right] \varphi_{zy}^{(4)} \varphi_{xy}^{(4)} d\ell \right\} s$$

$$A_{9,15} = \left\{ \frac{\rho U_0^2 S_2}{2\mathcal{M}_{xy}^{(4)}} \int_0^L \left[\left(\frac{\partial C_y(\ell)}{\partial p}\right)_0^e + \left(\frac{\partial C_D(\ell)}{\partial p}\right)_0^e \beta_0 \right] \varphi_{zy}^{(5)} \varphi_{xy}^{(4)} d\ell \right\} s$$

$$A_{9,16} = \left\{ \frac{\rho U_0^2 S_2}{2\mathcal{M}_{xy}^{(4)}} \int_0^L \left[\left(\frac{\partial C_y(\ell)}{\partial u}\right)_0^e + \left(\frac{\partial C_D(\ell)}{\partial u}\right)_0^e \beta_0 + \frac{2}{U_0} C_y(\ell)_0^e + \frac{2}{U_0} \beta_0 C_D(\ell)_0^e \right] \varphi_{xy}^{(4)} d\ell \right\}.$$

$$A_{9,17} = \left\{ \frac{M_R}{\mathcal{M}_{xy}^{(4)}} \varphi_{xy}^{(4)}(eG) + \frac{\rho U_0^2 S_2}{2\mathcal{M}_{xy}^{(4)}} \int_0^L \left[\left(\frac{\partial C_y(\ell)}{\partial \dot{v}}\right)_0^e + \left(\frac{\partial C_D(\ell)}{\partial \dot{v}}\right)_0^e \beta_0 \right] \varphi_{xy}^{(4)} d\ell \right\} s.$$

$$+ \left\{ \frac{\rho U_0^2 S_2}{2\mathcal{M}_{xy}^{(4)}} \int_0^L \left[\left(\frac{\partial C_y(\ell)}{\partial v}\right)_0^e + \left(\frac{\partial C_D(\ell)}{\partial v}\right)_0^e \beta_0 + \frac{1}{U_0} C_D(\ell)_0^e \right] \varphi_{xy}^{(4)} d\ell \right\}$$

$$A_{9,18} = \left\{ \frac{\rho U_0^2 S_2}{2\mathcal{M}_{xy}^{(4)}} \int_0^L \left[\left(\frac{\partial C_y(\ell)}{\partial w}\right)_0^e + \left(\frac{\partial C_D(\ell)}{\partial w}\right)_0^e \beta_0 \right] \varphi_{xy}^{(4)} d\ell \right\}.$$

$$A_{9,19} = \left\{ \frac{\rho U_0^2 S_2}{2\mathcal{M}_{xy}^{(4)}} \int_0^L \left[\left(\frac{\partial C_y(\ell)}{\partial p}\right)_0^e + \left(\frac{\partial C_D(\ell)}{\partial p}\right)_0^e \beta_0 \right] \varphi_{xy}^{(4)} d\ell \right\} s$$

$$A_{9,20} = \left\{ \frac{\rho U_0^2 S_2}{2\mathcal{M}_{xy}^{(4)}} \int_0^L \left[\left(\frac{\partial C_y(\ell)}{\partial q}\right)_0^e + \left(\frac{\partial C_D(\ell)}{\partial q}\right)_0^e \beta_0 \right] \varphi_{xy}^{(4)} d\ell \right\} s$$

$$A_{9,21} = \left\{ \frac{-M_R}{\mathcal{M}_{xy}^{(4)}} (L - \ell_{CG}) \varphi_{xy}^{(4)}(\ell_R) \right\} s^2 + \left\{ \frac{\rho U_0^2 S_2}{2\mathcal{M}_{xy}^{(4)}} \int_0^L \left[\left(\frac{\partial C_y(\ell)}{\partial r}\right)_0^e + \left(\frac{\partial C_D(\ell)}{\partial r}\right)_0^e \beta_0 \right] \varphi_{xy}^{(4)} d\ell \right\} s$$

$A_{9,22} \cdots A_{9,25} = 0.$

$$A_{9,26} = \left\{ \frac{M_{C_1}}{\mathcal{M}_{xy}^{(4)}} \sigma_{xy}^{(4)}(X_{1B}) \right\} s^2 + \left\{ \frac{K_1}{\mathcal{M}_{xy}^{(4)}} \varphi_{xy}^{(4)}(X_1) + \frac{M_{I_1}(T_s + T_c - D)}{M_t\, \mathcal{M}_{xy}^{(4)}} \sigma_{xy}^{(4)}(X_{1B}) \right\}$$

$$A_{9,27} = \left\{ \frac{M_{C_2}}{\mathcal{M}_{xy}^{(4)}} \sigma_{xy}^{(4)}(X_{2B}) \right\} s^2 + \left\{ \frac{K_2}{\mathcal{M}_{xy}^{(4)}} \varphi_{xy}^{(4)}(X_2) + \frac{M_{I_2}(T_s + T_c - D)}{M_t\, \mathcal{M}_{xy}^{(4)}} \sigma_{xy}^{(4)}(X_{2B}) \right\}$$

$$A_{9,28} = \left\{ \frac{M_{C_3}}{\mathcal{M}_{xy}^{(4)}} \sigma_{xy}^{(4)}(X_{3B}) \right\} s^2 + \left\{ \frac{K_3}{\mathcal{M}_{xy}^{(4)}} \varphi_{xy}^{(4)}(X_3) + \frac{M_{I_3}(T_s + T_c - D)}{M_t\, \mathcal{M}_{xy}^{(4)}} \sigma_{xy}^{(4)}(X_{3B}) \right\}$$

$$A_{9,29} = \left\{ \frac{M_{C_4}}{\mathcal{M}_{xy}^{(4)}} \sigma_{xy}^{(4)}(X_{4B}) \right\} s^2 + \left\{ \frac{K_4}{\mathcal{M}_{xy}^{(4)}} \varphi_{xy}^{(4)}(X_4) + \frac{M_{I_4}(T_s + T_c - D)}{M_t\, \mathcal{M}_{xy}^{(4)}} \sigma_{xy}^{(4)}(X_{4B}) \right\}$$

$A_{9,30} = 0.$

$$A_{9,31} = \left\{ \frac{M_R\, \ell_R}{\mathcal{M}_{xy}^{(4)}} \varphi_{xy}^{(4)}(\ell_R) \right\} s^2 + \left\{ \frac{T_c}{\mathcal{M}_{xy}^{(4)}} \varphi_{xy}^{(4)}(eG) \right\}.$$

$A_{9,32} \cdots A_{9,40} = 0.$

$$A_{10,1} = \left\{ \frac{\rho U_0^2 S_2}{2\mathcal{M}_{xy}^{(5)}} \int_0^L \left[\left(\frac{\partial C_y(\ell)}{\partial w} \right)_0^e + \left(\frac{\partial C_D(\ell)}{\partial w} \right)_0^e \beta_0 \right] \varphi_{xz}^{(1)} \varphi_{xy}^{(5)}\, d\ell \right\} s$$

$$+ \left\{ \frac{-\rho U_0^3 S_2}{2\mathcal{M}_{xy}^{(5)}} \int_0^L \left[\left(\frac{\partial C_y(\ell)}{\partial w} \right)_0^e + \left(\frac{\partial C_D(\ell)}{\partial w} \right)_0^e \beta_0 \right] \sigma_{xz}^{(1)} \varphi_{xy}^{(5)}\, d\ell \right\}.$$

$$A_{10,2} = \left\{ \frac{\rho U_0^2 S_2}{2\mathcal{M}_{xy}^{(5)}} \int_0^L \left[\left(\frac{\partial C_y(\ell)}{\partial w} \right)_0^e + \left(\frac{\partial C_D(\ell)}{\partial w} \right)_0^e \beta_0 \right] \varphi_{xz}^{(2)} \varphi_{xy}^{(5)}\, d\ell \right\} s$$

$$+ \left\{ \frac{-\rho U_0^3 S_2}{2\mathcal{M}_{xy}^{(5)}} \int_0^L \left[\left(\frac{\partial C_y(\ell)}{\partial w} \right)_0^e + \left(\frac{\partial C_D(\ell)}{\partial w} \right)_0^e \beta_0 \right] \sigma_{xz}^{(2)} \varphi_{xy}^{(5)}\, d\ell \right\}.$$

$$A_{10'3} = \left\{ \frac{\rho U_0^2 S_2}{2 \mathcal{M}_{xy}^{(5)}} \int_0^L \left[\left(\frac{\partial C_y(\ell)}{\partial w} \right)_0^e + \left(\frac{\partial C_D(\ell)}{\partial w} \right)_0^e \beta_0 \right] \varphi_{xz}^{(3)} \varphi_{xy}^{(5)} d\ell \right\} s$$

$$+ \left\{ \frac{-\rho U_0^3 S_2}{2 \mathcal{M}_{xy}^{(5)}} \int_0^L \left[\left(\frac{\partial C_y(\ell)}{\partial w} \right)_0^e + \left(\frac{\partial C_D(\ell)}{\partial w} \right)_0^e \beta_0 \right] \sigma_{xz}^{(3)} \varphi_{xy}^{(5)} d\ell \right\} .$$

$$A_{10'4} = \left\{ \frac{\rho U_0^2 S_2}{2 \mathcal{M}_{xy}^{(5)}} \int_0^L \left[\left(\frac{\partial C_y(\ell)}{\partial w} \right)_0^e + \left(\frac{\partial C_D(\ell)}{\partial w} \right)_0^e \beta_0 \right] \varphi_{xz}^{(4)} \varphi_{xy}^{(5)} d\ell \right\} s$$

$$+ \left\{ \frac{-\rho U_0^3 S_2}{2 \mathcal{M}_{xy}^{(5)}} \int_0^L \left[\left(\frac{\partial C_y(\ell)}{\partial w} \right)_0^e + \left(\frac{\partial C_D(\ell)}{\partial w} \right)_0^e \beta_0 \right] \sigma_{xz}^{(4)} \varphi_{xy}^{(5)} d\ell \right\} .$$

$$A_{10'5} = \left\{ \frac{\rho U_0^2 S_2}{2 \mathcal{M}_{xy}^{(5)}} \int_0^L \left[\left(\frac{\partial C_y(\ell)}{\partial w} \right)_0^e + \left(\frac{\partial C_D(\ell)}{\partial w} \right)_0^e \beta_0 \right] \varphi_{xz}^{(5)} \varphi_{xy}^{(5)} d\ell \right\} s$$

$$+ \left\{ \frac{-\rho U_0^3 S_2}{2 \mathcal{M}_{xy}^{(5)}} \int_0^L \left[\left(\frac{\partial C_y(\ell)}{\partial w} \right)_0^e + \left(\frac{\partial C_D(\ell)}{\partial w} \right)_0^e \beta_0 \right] \sigma_{xz}^{(5)} \varphi_{xy}^{(5)} d\ell \right\} .$$

$$A_{10,6} = \left\{ \frac{M_R}{\mathcal{M}_{xy}^{(5)}} \left(\varphi_{xy}^{(1)}(eG) - \ell_R \sigma_{xy}^{(1)}(eG) \right) \varphi_{xy}^{(5)}(\ell_R) \right\} s^2 + \left\{ \frac{\rho U_0^2 S_2}{2 \mathcal{M}_{xy}^{(5)}} \int_0^L \left[\left(\frac{\partial C_y(\ell)}{\partial v} \right)_0^e + \left(\frac{\partial C_D(\ell)}{\partial v} \right)_0^e \beta_0 \right.\right.$$

$$\left.\left. + \frac{1}{U_0} C_D(\ell)_0^e \right] \varphi_{xy}^{(1)} \varphi_{xy}^{(5)} d\ell \right\} s + \left\{ - \frac{(T_s + T_c)}{\mathcal{M}_{xy}^{(5)}} \sigma_{xy}^{(1)}(eG) \varphi_{xy}^{(5)}(eG) \right.$$

$$\left. - \frac{\rho U_0^3 S_2}{2 \mathcal{M}_{xy}^{(5)}} \int_0^L \left[\left(\frac{\partial C_y(\ell)}{\partial v} \right)_0^e + \left(\frac{\partial C_D(\ell)}{\partial v} \right)_0^e \beta_0 + \frac{1}{U_0} C_D(\ell)_0^e \right] \sigma_{xy}^{(1)} \varphi_{xy}^{(5)} d\ell \right\} .$$

$$A_{10,7} = \left\{ \frac{M_R}{\mathcal{M}_{xy}^{(5)}} \left(\varphi_{xy}^{(2)}(eG) - \ell_R \sigma_{xy}^{(2)}(eG) \right) \varphi_{xy}^{(5)}(\ell_R) \right\} s^2 + \left\{ \frac{\rho U_0^2 S_2}{2 \mathcal{M}_{xy}^{(5)}} \int_0^L \left[\left(\frac{\partial C_y(\ell)}{\partial v} \right)_0^e + \left(\frac{\partial C_D(\ell)}{\partial v} \right)_0^e \beta_0 \right.\right.$$

$$\left.\left. + \frac{1}{U_0} C_D(\ell)_0^e \right] \varphi_{xy}^{(2)} \varphi_{xy}^{(5)} d\ell \right\} s + \left\{ - \frac{(T_s + T_c)}{\mathcal{M}_{xy}^{(\,)}} \sigma_{xy}^{(2)}(eG) \varphi_{xy}^{(5)}(eG) \right.$$

$$\left. - \frac{\rho U_0^3 S_2}{2 \mathcal{M}_{xy}^{(5)}} \int_0^L \left[\left(\frac{\partial C_y(\ell)}{\partial v} \right)_0^e + \left(\frac{\partial C_D(\ell)}{\partial v} \right) \beta_0 + \frac{1}{U_0} C_D(\ell)_0^e \right] \sigma_{xy}^{(2)} \varphi_{xy}^{(5)} d\ell \right\} .$$

$$A_{10,8} = \left\{ \frac{M_R}{\mathcal{M}^{(5)}_{xy}} \left(\varphi^{(3)}_{xy}(eG) - \ell_R \sigma^{(3)}_{xy}(eG) \right) \varphi^{(5)}_{xy}(\ell_R) \right\} s^2 + \left\{ \frac{\rho U_0^2 S_2}{2\mathcal{M}^{(5)}_{xy}} \int_0^L \left[\left(\frac{\partial C_y(\ell)}{\partial v} \right)_0^e + \left(\frac{\partial C_D(\ell)}{\partial v} \right)_0^e \beta_0 \right.\right.$$

$$\left.\left. + \frac{1}{U_0} C_D(\ell)_0^e \right] \varphi^{(3)}_{xy} \varphi^{(5)}_{xy} d\ell \right\} s + \left\{ - \frac{(T_s + T_c)}{\mathcal{M}^{(5)}_{xy}} \sigma^{(3)}_{xy}(eG)\, \varphi^{(5)}_{xy}(eG) \right.$$

$$\left. - \frac{\rho U_0^3 S_2}{2\mathcal{M}^{(5)}_{xy}} \int_0^L \left[\left(\frac{\partial C_y(\ell)}{\partial v} \right)_0^e + \left(\frac{\partial C_D(\ell)}{\partial v} \right) \beta_0 + \frac{1}{U_0} C_D(\ell)_0^e \right] \sigma^{(3)}_{xy} \varphi^{(5)}_{xy} d\ell \right\}.$$

$$A_{10,9} = \left\{ \frac{M_R}{\mathcal{M}^{(5)}_{xy}} \left(\varphi^{(4)}_{xy}(eG) - \ell_R \sigma^{(4)}_{xy}(eG) \right) \varphi^{(5)}_{xy}(\ell_R) \right\} s^2 + \left\{ \frac{\rho U_0^2 S_2}{2\mathcal{M}^{(5)}_{xy}} \int_0^L \left[\left(\frac{\partial C_y(\ell)}{\partial v} \right)_0^e + \left(\frac{\partial C_D(\ell)}{\partial v} \right)_0^e \beta_0 \right.\right.$$

$$\left.\left. + \frac{1}{U_0} C_D(\ell)_0^e \right] \varphi^{(4)}_{xy} \varphi^{(5)}_{xy} d\ell \right\} s + \left\{ - \frac{(T_s + T_c)}{\mathcal{M}^{(5)}_{xy}} \sigma^{(4)}_{xy}(eG)\, \varphi^{(5)}_{xy}(eG) \right.$$

$$\left. - \frac{\rho U_0^3 S_2}{2\mathcal{M}^{(5)}_{xy}} \int_0^L \left[\left(\frac{\partial C_y(\ell)}{\partial v} \right)_0^e + \left(\frac{\partial C_D(\ell)}{\partial v} \right)_0^e \beta_0 + \frac{1}{U_0} C_D(\ell)_0^e \right] \sigma^{(4)}_{xy} \varphi^{(5)}_{xy} d\ell \right\}.$$

$$A_{10,10} = \left\{ 1 + \frac{M_R}{\mathcal{M}^{(5)}_{xy}} \left(\varphi^{(5)}_{xy}(eG) - \ell_R \sigma^{(5)}_{xy}(eG) \right) \varphi^{(5)}_{xy}(\ell_R) \right\} s^2 + \left\{ 2\xi^{(5)}_{xy} \omega^{(5)}_{xy} \right.$$

$$\left. + \frac{\rho U_0^2 S_2}{2\mathcal{M}^{(5)}_{xy}} \int_0^L \left[\left(\frac{\partial C_y(\ell)}{\partial v} \right)_0^e + \left(\frac{\partial C_D(\ell)}{\partial v} \right)_0^e \beta_0 + \frac{1}{U_0} C_D(\ell)_0^e \right] \varphi^{(5)2}_{xy} d\ell \right\} s + \left\{ \omega^{(5)2}_{xy} \right.$$

$$- \frac{(T_s + T_c)}{\mathcal{M}^{(5)}_{xy}} \sigma^{(5)}_{xy}(eG)\, \varphi^{(5)}_{xy}(eG) - \frac{\rho U_0^3 S_2}{2\mathcal{M}^{(5)}_{xy}} \int_0^L \left[\left(\frac{\partial C_y(\ell)}{\partial v} \right)_0^e + \left(\frac{\partial C_D(\ell)}{\partial v} \right)_0^e \beta_0 \right.$$

$$\left.\left. + \frac{1}{U_0} C_D(\ell)_0^e \right] \sigma^{(5)}_{xy} \varphi^{(5)}_{xy} d\ell \right\}.$$

$$A_{10,11} = \left\{ \frac{\rho U_0^2 S_2}{2\mathcal{M}^{(5)}_{xy}} \int_0^L \left[\left(\frac{\partial C_y(\ell)}{\partial p} \right)_0^e + \left(\frac{\partial C_D(\ell)}{\partial p} \right)_0^e \beta_0 \right] \varphi^{(1)}_{zy} \varphi^{(5)}_{xy} d\ell \right\} s$$

$$A_{10,12} = \left\{ \frac{\rho U_0^2 S_2}{2\mathcal{M}^{(5)}_{xy}} \int_0^L \left[\left(\frac{\partial C_y(\ell)}{\partial p} \right)_0^e + \left(\frac{\partial C_D(\ell)}{\partial p} \right)_0^e \beta_0 \right] \varphi^{(2)}_{zy} \varphi^{(5)}_{xy} d\ell \right\} s$$

$$A_{10,13} = \left\{ \frac{\rho U_0^2 S_2}{2\mathcal{M}_{xy}^{(5)}} \int_0^L \left[\left(\frac{\partial C_y(\ell)}{\partial p} \right)_0^e + \left(\frac{\partial C_D(\ell)}{\partial p} \right)_0^e \beta_0 \right] \varphi_{zy}^{(3)} \varphi_{xy}^{(5)} d\ell \right\} s.$$

$$A_{10,14} = \left\{ \frac{\rho U_0^2 S_2}{2\mathcal{M}_{xy}^{(5)}} \int_0^L \left[\left(\frac{\partial C_y(\ell)}{\partial p} \right)_0^e + \left(\frac{\partial C_D(\ell)}{\partial p} \right)_0^e \beta_0 \right] \varphi_{zy}^{(4)} \varphi_{xy}^{(5)} d\ell \right\} s.$$

$$A_{10,15} = \left\{ \frac{\rho U_0^2 S_2}{2\mathcal{M}_{xy}^{(5)}} \int_0^L \left[\left(\frac{\partial C_y(\ell)}{\partial p} \right)_0^e + \left(\frac{\partial C_D(\ell)}{\partial p} \right)_0^e \beta_0 \right] \varphi_{zy}^{(5)} \varphi_{xy}^{(5)} d\ell \right\} s.$$

$$A_{10,16} = \left\{ \frac{\rho U_0^2 S_2}{2\mathcal{M}_{xy}^{(5)}} \int_0^L \left[\left(\frac{\partial C_y(\ell)}{\partial u} \right)_0^e + \left(\frac{\partial C_D(\ell)}{\partial u} \right)_0^e \beta_0 + \frac{2}{U_0} C_y(\ell)_0^e + \frac{2}{U_0} \beta_0 C_D(\ell)_0^e \right] \varphi_{xy}^{(5)} d\ell \right\}.$$

$$A_{10,17} = \left\{ \frac{M_R}{\mathcal{M}_{xy}^{(5)}} \varphi_{xy}^{(5)}(\ell_R) + \frac{\rho U_0^2 S_2}{2\mathcal{M}_{xy}^{(5)}} \int_0^L \left[\left(\frac{\partial C_y(\ell)}{\partial \dot{v}} \right)_0^e + \left(\frac{\partial C_D(\ell)}{\partial \dot{v}} \right)_0^e \beta_0 \right] \varphi_{xy}^{(5)} d\ell \right\} s$$

$$+ \left\{ \frac{\rho U_0^2 S_2}{2\mathcal{M}_{xy}^{(5)}} \int_0^L \left[\left(\frac{\partial C_y(\ell)}{\partial v} \right)_0^e + \left(\frac{\partial C_D(\ell)}{\partial v} \right)_0^e \beta_0 + \frac{1}{U_0} C_D(\ell)_0^e \right] \varphi_{xy}^{(5)} d\ell \right\}$$

$$A_{10,18} = \left\{ \frac{\rho U_0^2 S_2}{2\mathcal{M}_{xy}^{(5)}} \int_0^L \left[\left(\frac{\partial C_y(\ell)}{\partial w} \right)_0^e + \left(\frac{\partial C_D(\ell)}{\partial w} \right)_0^e \beta_0 \right] \varphi_{xy}^{(5)} d\ell \right\}$$

$$A_{10,19} = \left\{ \frac{\rho U_0^2 S_2}{2\mathcal{M}_{xy}^{(5)}} \int_0^L \left[\left(\frac{\partial C_y(\ell)}{\partial p} \right)_0^e + \left(\frac{\partial C_D(\ell)}{\partial p} \right)_0^e \beta_0 \right] \varphi_{xy}^{(5)} d\ell \right\} s.$$

$$A_{10,20} = \left\{ \frac{\rho U_0^2 S_2}{2\mathcal{M}_{xy}^{(5)}} \int_0^L \left[\left(\frac{\partial C_y(\ell)}{\partial q} \right)_0^e + \left(\frac{\partial C_D(\ell)}{\partial q} \right)_0^e \beta_0 \right] \varphi_{xy}^{(5)} d\ell \right\} s$$

$$A_{10,21} = \left\{ \frac{-M_R}{\mathcal{M}_{xy}^{(5)}} (L - \ell_{CG}) \varphi_{xy}^{(5)}(\ell_R) \right\} s^2 + \left\{ \frac{\rho U_0^2 S_2}{2\mathcal{M}_{xy}^{(5)}} \int_0^L \left(\frac{\partial C_y(\ell)}{\partial r} \right)_0^e + \left(\frac{\partial C_D(\ell)}{\partial r} \right)_0^e \beta_0 \right] \varphi_{xy}^{(5)} d\ell \right\} s.$$

$$A_{10,22} \cdots A_{10,25} = 0.$$

$$A_{10,26} = \left\{ \frac{M_{C_1}}{\mathcal{M}_{xy}^{(5)}} \sigma_{xy}^{(5)}(X_{1B}) \right\} s^2 + \left\{ \frac{K_1}{\mathcal{M}_{xy}^{(5)}} \varphi_{xy}^{(5)}(X_1) + \frac{M_{I_1}(T_s + T_c - D)}{M_t \mathcal{M}_{xy}^{(5)}} \sigma_{xy}^{(5)}(X_{1B}) \right\}$$

$$A_{10,27} = \left\{ \frac{M_{C_2}}{\mathcal{M}_{xy}^{(5)}} \sigma_{xy}^{(5)}(X_{2B}) \right\} s^2 + \left\{ \frac{K_2}{\mathcal{M}_{xy}^{(5)}} \varphi_{xy}^{(5)}(X_2) + \frac{M_{I_2}(T_s + T_c - D)}{M_t \mathcal{M}_{xy}^{(5)}} \sigma_{xy}^{(5)}(X_{2B}) \right\}$$

$$A_{10,28} = \left\{ \frac{M_{C_3}}{\mathcal{M}_{xy}^{(5)}} \sigma_{xy}^{(5)}(X_{3B}) \right\} s^2 + \left\{ \frac{K_3}{\mathcal{M}_{xy}^{(5)}} \varphi_{xy}^{(5)}(X_3) + \frac{M_{I_3}(T_s + T_c - D)}{M_t \mathcal{M}_{xy}^{(5)}} \sigma_{xy}^{(5)}(X_{3B}) \right\}$$

$$A_{10,29} = \left\{ \frac{M_{C_4}}{\mathcal{M}_{xy}^{(5)}} \sigma_{xy}^{(5)}(X_{4B}) \right\} s^2 + \left\{ \frac{K_4}{\mathcal{M}_{xy}^{(5)}} \varphi_{xy}^{(5)}(X_4) + \frac{M_{I_4}(T_s + T_c - D)}{M_t \mathcal{M}_{xy}^{(5)}} \sigma_{xy}^{(5)}(X_{4B}) \right\}$$

$$A_{10,30} = 0.$$

$$A_{10,31} = \left\{ \frac{M_R \ell_R}{\mathcal{M}_{xy}^{(5)}} \varphi_{xy}^{(5)}(\ell_R) \right\} s^2 + \left\{ \frac{T_c}{\mathcal{M}_{xy}^{(5)}} \varphi_{xy}^{(5)}(eG) \right\}.$$

$$A_{10,32} \;\cdots\; A_{10,40} = 0.$$

$$A_{11,1} = \left\{ \frac{-\rho U_0^2 S_4 b_1}{2\mathcal{M}_{zy}^{(1)}} \int_0^L \left(\frac{\partial c_\ell(\ell)}{\partial w} \right)_0^e \varphi_{xz}^{(1)} \varphi_{zy}^{(1)} \, d\ell \right\} s$$
$$+ \left\{ \frac{\rho U_0^3 S_4 b_1}{2\mathcal{M}_{zy}^{(1)}} \int_0^L \left(\frac{\partial c_\ell(\ell)}{\partial w} \right)_0^e \sigma_{xz}^{(1)} \varphi_{zy}^{(1)} \, d\ell \right\}.$$

$$A_{11,2} = \left\{ \frac{-\rho U_0^2 S_4 b_1}{2\mathcal{M}_{zy}^{(1)}} \int_0^L \left(\frac{\partial c_\ell(\ell)}{\partial w} \right)_0^e \varphi_{xz}^{(2)} \varphi_{zy}^{(1)} \, d\ell \right\} s$$
$$+ \left\{ \frac{\rho U_0^3 S_4 b_1}{2\mathcal{M}_{zy}^{(1)}} \int_0^L \left(\frac{\partial c_\ell(\ell)}{\partial w} \right)_0^e \sigma_{xz}^{(2)} \varphi_{zy}^{(1)} \, d\ell \right\}.$$

$$A_{11,3} = \left\{ \frac{-\rho U_0^2 S_4 b_1}{2\mathcal{M}_{zy}^{(1)}} \int_0^L \left(\frac{\partial c_\ell(\ell)}{\partial w} \right)_0^e \varphi_{xz}^{(3)} \varphi_{zy}^{(1)} \, d\ell \right\} s$$
$$+ \left\{ \frac{\rho U_0^3 S_4 b_1}{2\mathcal{M}_{zy}^{(1)}} \int_0^L \left(\frac{\partial c_\ell(\ell)}{\partial w} \right)_0^e \sigma_{xz}^{(3)} \varphi_{zy}^{(1)} \, d\ell \right\}.$$

$$A_{11,4} = \left\{ \frac{-\rho U_0^2 S_4 b_1}{2\mathcal{M}_{zy}^{(1)}} \int_0^L \left(\frac{\partial c_\ell(\ell)}{\partial w} \right)_0^e \varphi_{xz}^{(4)} \varphi_{zy}^{(1)} \, d\ell \right\} s$$
$$+ \left\{ \frac{\rho U_0^3 S_4 b_1}{2\mathcal{M}_{zy}^{(1)}} \int_0^L \left(\frac{\partial c_\ell(\ell)}{\partial w} \right)_0^e \sigma_{xz}^{(4)} \varphi_{zy}^{(1)} \, d\ell \right\}.$$

$$A_{11,5} = \left\{ \frac{-\rho U_0^2 S_4 b_1}{2\mathcal{M}_{zy}^{(1)}} \int_0^L \left(\frac{\partial C_\ell(\ell)}{\partial w} \right)_0^e \varphi_{xz}^{(5)} \varphi_{zy}^{(1)} d\ell \right\} s + \left\{ \frac{\rho U_0^3 S_4 b_1}{2\mathcal{M}_{zy}^{(1)}} \int_0^L \left(\frac{\partial C_\ell(\ell)}{\partial w} \right)_0^e \sigma_{xz}^{(5)} \varphi_{zy}^{(1)} d\ell \right\},$$

$$A_{11,6} = \left\{ \frac{-\rho U_0^2 S_4 b_1}{2\mathcal{M}_{zy}^{(1)}} \int_0^L \left(\frac{\partial C_\ell(\ell)}{\partial v} \right)_0^e \varphi_{xy}^{(1)} \varphi_{zy}^{(1)} d\ell \right\} s + \left\{ \frac{\rho U_0^3 S_4 b_1}{2\mathcal{M}_{zy}^{(1)}} \int_0^L \left(\frac{\partial C_\ell(\ell)}{\partial v} \right)_0^e \sigma_{xy}^{(1)} \varphi_{zy}^{(1)} d\ell \right\},$$

$$A_{11,7} = \left\{ \frac{-\rho U_0^2 S_4 b_1}{2\mathcal{M}_{zy}^{(1)}} \int_0^L \left(\frac{\partial C_\ell(\ell)}{\partial v} \right)_0^e \varphi_{xy}^{(2)} \varphi_{zy}^{(1)} d\ell \right\} s + \left\{ \frac{\rho U_0^3 S_4 b_1}{2\mathcal{M}_{zy}^{(1)}} \int_0^L \left(\frac{\partial C_\ell(\ell)}{\partial v} \right)_0^e \sigma_{xy}^{(2)} \varphi_{zy}^{(1)} d\ell \right\},$$

$$A_{11,8} = \left\{ \frac{-\rho U_0^2 S_4 b_1}{2\mathcal{M}_{zy}^{(1)}} \int_0^L \left(\frac{\partial C_\ell(\ell)}{\partial v} \right)_0^e \varphi_{xy}^{(3)} \varphi_{zy}^{(1)} d\ell \right\} s + \left\{ \frac{\rho U_0^3 S_4 b_1}{2\mathcal{M}_{zy}^{(1)}} \int_0^L \left(\frac{\partial C_\ell(\ell)}{\partial v} \right)_0^e \sigma_{xy}^{(3)} \varphi_{zy}^{(1)} d\ell \right\},$$

$$A_{11,9} = \left\{ \frac{-\rho U_0^2 S_4 b_1}{2\mathcal{M}_{zy}^{(1)}} \int_0^L \left(\frac{\partial C_\ell(\ell)}{\partial v} \right)_0^e \varphi_{xy}^{(4)} \varphi_{zy}^{(1)} d\ell \right\} s + \left\{ \frac{\rho U_0^3 S_4 b_1}{2\mathcal{M}_{zy}^{(1)}} \int_0^L \left(\frac{\partial C_\ell(\ell)}{\partial v} \right)_0^e \sigma_{xy}^{(4)} \varphi_{zy}^{(1)} d\ell \right\},$$

$$A_{11,10} = \left\{ \frac{-\rho U_0^2 S_4 b_1}{2\mathcal{M}_{zy}^{(1)}} \int_0^L \left(\frac{\partial C_\ell(\ell)}{\partial v} \right)_0^e \varphi_{xy}^{(5)} \varphi_{zy}^{(1)} d\ell \right\} s + \left\{ \frac{\rho U_0^3 S b_1}{2\mathcal{M}_{zy}^{(1)}} \int_0^L \left(\frac{\partial C_\ell(\ell)}{\partial v} \right)_0^e \sigma_{xy}^{(5)} \varphi_{zy}^{(1)} d\ell \right\},$$

$$A_{11,11} = \{1\} s^2 + \left\{ 2\xi_{zy}^{(1)} \omega_{zy}^{(1)} - \frac{\rho U_0^2 S_4 b_1}{2\mathcal{M}_{zy}^{(1)}} \int_0^L \left(\frac{\partial C_\ell(\ell)}{\partial p} \right)_0^e \varphi_{zy}^{(1)2} d\ell \right\} s + \left\{ \omega_{zy}^{(1)2} \right\}$$

$$A_{11,12} = \left\{ \frac{-\rho U_0^2 S_4 b_1}{2\mathcal{M}_{zy}^{(1)}} \int_0^L \left(\frac{\partial C_\ell(\ell)}{\partial p} \right)_0^e \varphi_{zy}^{(2)} \varphi_{zy}^{(1)} d\ell \right\} s.$$

$$A_{11',13} = \left\{ \frac{-\rho U_0^2 S_4 b_1}{2 \mathcal{M}_{zy}^{(1)}} \int_0^L \left(\frac{\partial C_{\ell}(\ell)}{\partial p} \right)_0^e \varphi_{zy}^{(3)} \varphi_{zy}^{(1)} d\ell \right\} s.$$

$$A_{11',14} = \left\{ \frac{-\rho U_0^2 S_4 b_1}{2 \mathcal{M}_{zy}^{(1)}} \int_0^L \left(\frac{\partial C_{\ell}(\ell)}{\partial p} \right)_0^e \varphi_{zy}^{(4)} \varphi_{zy}^{(1)} d\ell \right\} s.$$

$$A_{11',15} = \left\{ \frac{-\rho U_0^2 S_4 b_1}{2 \mathcal{M}_{zy}^{(1)}} \int_0^L \left(\frac{\partial C_{\ell}(\ell)}{\partial p} \right)_0^e \varphi_{zy}^{(5)} \varphi_{zy}^{(1)} d\ell \right\} s.$$

$$A_{11',16} = \left\{ \frac{-\rho U_0^2 S_4 b_1}{2 \mathcal{M}_{zy}^{(1)}} \int_0^L \left[\left(\frac{\partial C_{\ell}(\ell)}{\partial u} \right)_0^e + \frac{2}{U_0} C_{\ell}(\ell)_0^e \right] \varphi_{zy}^{(1)} d\ell \right\}.$$

$$A_{11',17} = \left\{ \frac{-\rho U_0^2 S_4 b_1}{2 \mathcal{M}_{zy}^{(1)}} \int_0^L \left(\frac{\partial C_{\ell}(\ell)}{\partial v} \right)_0^e \varphi_{zy}^{(1)} d\ell \right\}.$$

$$A_{11',18} = \left\{ \frac{-\rho U_0^2 S_4 b_1}{2 \mathcal{M}_{zy}^{(1)}} \int_0^L \left(\frac{\partial C_{\ell}(\ell)}{\partial w} \right)_0^e \varphi_{zy}^{(1)} d\ell \right\}.$$

$$A_{11',19} = \left\{ \frac{-\rho U_0^2 S_4 b_1}{2 \mathcal{M}_{zy}^{(1)}} \int_0^L \left(\frac{\partial C_{\ell}(\ell)}{\partial p} \right)_0^e \varphi_{zy}^{(1)} d\ell \right\} s$$

$$A_{11',20} = \left\{ \frac{-\rho U_0^2 S_4 b_1}{2 \mathcal{M}_{zy}^{(1)}} \int_0^L \left(\frac{\partial C_{\ell}(\ell)}{\partial q} \right)_0^e \varphi_{zy}^{(1)} d\ell \right\} s$$

$$A_{11',21} = \left\{ \frac{-\rho U_0^2 S_4 b_1}{2 \mathcal{M}_{zy}^{(1)}} \int_0^L \left(\frac{\partial C_{\ell}(\ell)}{\partial r} \right)_0^e \varphi_{zy}^{(1)} d\ell \right\} s$$

$$A_{11',22} \cdots\cdots A_{11',40} = 0.$$

$$A_{12',1} = \left\{ \frac{-\rho U_0^2 S_4 b_1}{2 \mathcal{M}_{zy}^{(2)}} \int_0^L \left(\frac{\partial C_{\ell}(\ell)}{\partial w} \right)_0^e \varphi_{xz}^{(1)} \varphi_{zy}^{(2)} d\ell \right\} s$$
$$+ \left\{ \frac{\rho U_0^3 S_4 b_1}{2 \mathcal{M}_{zy}^{(2)}} \int_0^L \left(\frac{\partial C_{\ell}(\ell)}{\partial w} \right)_0^e \sigma_{xz}^{(1)} \varphi_{zy}^{(2)} d\ell \right\}.$$

$$A_{12,2} = \left\{ \frac{-\rho U_0^2 S_4 b_1}{2 \mathcal{M}_{zy}^{(2)}} \int_0^L \left(\frac{\partial C_\ell(\ell)}{\partial w} \right)_0^e \varphi_{xz}^{(2)} \varphi_{zy}^{(2)} d\ell \right\} s$$

$$+ \left\{ \frac{\rho U_0^3 S_4 b_1}{2 \mathcal{M}_{zy}^{(2)}} \int_0^L \left(\frac{\partial C_\ell(\ell)}{\partial w} \right)_0^e \sigma_{xz}^{(2)} \varphi_{zy}^{(2)} d\ell \right\},$$

$$A_{12,3} = \left\{ \frac{-\rho U_0^2 S_4 b_1}{2 \mathcal{M}_{zy}^{(2)}} \int_0^L \left(\frac{\partial C_\ell(\ell)}{\partial w} \right)_0^e \varphi_{xz}^{(3)} \varphi_{zy}^{(2)} d\ell \right\} s$$

$$+ \left\{ \frac{\rho U_0^3 S_4 b_1}{2 \mathcal{M}_{zy}^{(2)}} \int_0^L \left(\frac{\partial C_\ell(\ell)}{\partial w} \right)_0^e \sigma_{xz}^{(3)} \varphi_{zy}^{(2)} d\ell \right\},$$

$$A_{12'4} = \left\{ \frac{-\rho U_0^2 S_4 b_1}{2 \mathcal{M}_{zy}^{(2)}} \int_0^L \left(\frac{\partial C_\ell(\ell)}{\partial w} \right)_0^e \varphi_{xz}^{(4)} \varphi_{zy}^{(2)} d\ell \right\} s$$

$$+ \left\{ \frac{\rho U_0^3 S_4 b_1}{2 \mathcal{M}_{zy}^{(2)}} \int_0^L \left(\frac{\partial C_\ell(\ell)}{\partial w} \right)_0^e \sigma_{xz}^{(4)} \varphi_{zy}^{(2)} d\ell \right\},$$

$$A_{12'5} = \left\{ \frac{-\rho U_0^2 S_4 b_1}{2 \mathcal{M}_{zy}^{(2)}} \int_0^L \left(\frac{\partial C_\ell(\ell)}{\partial w} \right)_0^e \varphi_{xz}^{(5)} \varphi_{zy}^{(2)} d\ell \right\} s$$

$$+ \left\{ \frac{\rho U_0^3 S_4 b_1}{2 \mathcal{M}_{zy}^{(2)}} \int_0^L \left(\frac{\partial C_\ell(\ell)}{\partial w} \right)_0^e \sigma_{xz}^{(5)} \varphi_{zy}^{(2)} d\ell \right\},$$

$$A_{12'6} = \left\{ \frac{-\rho U_0^2 S_4 b_1}{2 \mathcal{M}_{zy}^{(2)}} \int_0^L \left(\frac{\partial C_\ell(\ell)}{\partial v} \right)_0^e \varphi_{xy}^{(1)} \varphi_{zy}^{(2)} d\ell \right\} s$$

$$+ \left\{ \frac{\rho U_0^3 S_4 b_1}{2 \mathcal{M}_{zy}^{(2)}} \int_0^L \left(\frac{\partial C_\ell(\ell)}{\partial v} \right)_0^e \sigma_{xy}^{(1)} \varphi_{zy}^{(2)} d\ell \right\},$$

$$A_{12'7} = \left\{ \frac{-\rho U_0^2 S_4 b_1}{2 \mathcal{M}_{zy}^{(2)}} \int_0^L \left(\frac{\partial C_\ell(\ell)}{\partial v} \right)_0^e \varphi_{xy}^{(2)} \varphi_{zy}^{(2)} d\ell \right\} s$$

$$+ \left\{ \frac{\rho U_0^3 S_4 b_1}{2 \mathcal{M}_{zy}^{(2)}} \int_0^L \left(\frac{\partial C_\ell(\ell)}{\partial v} \right)_0^e \sigma_{xy}^{(2)} \varphi_{zy}^{(2)} d\ell \right\},$$

$$A_{12'8} = \left\{ \frac{-\rho U_0^2 S_4 b_1}{2 \mathcal{M}_{zy}^{(2)}} \int_0^L \left(\frac{\partial C_\ell(\ell)}{\partial v} \right)_0^e \varphi_{xy}^{(3)} \varphi_{zy}^{(2)} d\ell \right\} s$$

$$+ \left\{ \frac{\rho U_0^3 S_4 b_1}{2 \mathcal{M}_{zy}^{(2)}} \int_0^L \left(\frac{\partial C_\ell(\ell)}{\partial v} \right)_0^e \sigma_{xy}^{(3)} \varphi_{zy}^{(2)} d\ell \right\},$$

$$A_{12'9} = \left\{ \frac{-\rho U_0^2 S_4 b_1}{2 \mathcal{M}_{zy}^{(2)}} \int_0^L \left(\frac{\partial c_\ell(\ell)}{\partial v} \right)_0^e \varphi_{xy}^{(4)} \varphi_{zy}^{(2)} d\ell \right\} s + \left\{ \frac{\rho U_0^3 S_4 b_1}{2 \mathcal{M}_{zy}^{(2)}} \int_0^L \left(\frac{\partial c_\ell(\ell)}{\partial v} \right)_0^e \sigma_{xy}^{(4)} \varphi_{zy}^{(2)} d\ell \right\},$$

$$A_{12'10} = \left\{ \frac{-\rho U_0^2 S_4 b_1}{2 \mathcal{M}_{zy}^{(2)}} \int_0^L \left(\frac{\partial c_\ell(\ell)}{\partial v} \right)_0^e \varphi_{xy}^{(5)} \varphi_{zy}^{(2)} d\ell \right\} s + \left\{ \frac{\rho U_0^3 S_4 b_1}{2 \mathcal{M}_{zy}^{(2)}} \int_0^L \left(\frac{\partial c_\ell(\ell)}{\partial v} \right)_0^e \sigma_{xy}^{(5)} \varphi_{zy}^{(2)} d\ell \right\},$$

$$A_{12'11} = \left\{ \frac{-\rho U_0^2 S_4 b_1}{2 \mathcal{M}_{zy}^{(2)}} \int_0^L \left(\frac{\partial c_\ell(\ell)}{\partial p} \right)_0^e \varphi_{zy}^{(1)} \varphi_{zy}^{(2)} d\ell \right\} s.$$

$$A_{12'12} = \left\{ 1 \right\} s^2 + \left\{ 2\xi_{zy}^{(2)} \omega_{zy}^{(2)} - \frac{\rho U_0^2 S_4 b_1}{2 \mathcal{M}_{zy}^{(2)}} \int_0^L \left(\frac{\partial c_\ell(\ell)}{\partial p} \right)_0^e \varphi_{zy}^{(2)2} d\ell \right\} s + \left\{ \omega_{zy}^{(2)2} \right\}$$

$$A_{12'13} = \left\{ \frac{-\rho U_0^2 S_4 b_1}{2 \mathcal{M}_{zy}^{(2)}} \int_0^L \left(\frac{\partial c_\ell(\ell)}{\partial p} \right)_0^e \varphi_{zy}^{(3)} \varphi_{zy}^{(2)} d\ell \right\} s.$$

$$A_{12'14} = \left\{ \frac{-\rho U_0^2 S_4 b_1}{2 \mathcal{M}_{zy}^{(2)}} \int_0^L \left(\frac{\partial c_\ell(\ell)}{\partial p} \right)_0^e \varphi_{zy}^{(4)} \varphi_{zy}^{(2)} d\ell \right\} s.$$

$$A_{12'15} = \left\{ \frac{-\rho U_0^2 S_4 b_1}{2 \mathcal{M}_{zy}^{(2)}} \int_0^L \left(\frac{\partial c_\ell(\ell)}{\partial p} \right)_0^e \varphi_{zy}^{(5)} \varphi_{zy}^{(2)} d\ell \right\} s.$$

$$A_{12'16} = \left\{ \frac{-\rho U_0^2 S_4 b_1}{2 \mathcal{M}_{zy}^{(2)}} \int_0^L \left[\left(\frac{\partial c_\ell(\ell)}{\partial u} \right)_0^e + \frac{2}{U_0} c_\ell(\ell)_0^e \right] \varphi_{zy}^{(2)} d\ell \right\}.$$

$$A_{12'17} = \left\{ \frac{-\rho U_0^2 S_4 b_1}{2 \mathcal{M}_{zy}^{(2)}} \int_0^L \left(\frac{\partial c_\ell(\ell)}{\partial v} \right)_0^e \varphi_{zy}^{(2)} d\ell \right\}.$$

$$A_{12'18} = \left\{ \frac{-\rho U_0^2 S_4 b_1}{2 \mathcal{M}_{zy}^{(2)}} \int_0^L \left(\frac{\partial c_\ell(\ell)}{\partial w} \right)_0^e \varphi_{zy}^{(2)} d\ell \right\}.$$

$$A_{12'19} = \left\{ \frac{-\rho U_0^2 S_4 b_1}{2 \mathcal{M}_{zy}^{(2)}} \int_0^L \left(\frac{\partial c_\ell(\ell)}{\partial p} \right)_0^e \varphi_{zy}^{(2)} d\ell \right\} s$$

$$A_{12'20} = \left\{\frac{-\rho U_0^2 S_4 b_1}{2\mathcal{M}_{zy}^{(2)}}\int_0^L \left(\frac{\partial c_\ell(\ell)}{\partial q}\right)_0^e \varphi_{zy}^{(2)} d\ell\right\} s$$

$$A_{12'21} = \left\{\frac{-\rho U_0^2 S_4 b_1}{2\mathcal{M}_{zy}^{(2)}}\int_0^L \left(\frac{\partial c_\ell(\ell)}{\partial r}\right)_0^e \varphi_{zy}^{(2)} d\ell\right\} s$$

$$A_{12'22} \quad \cdots \quad A_{12'40} = 0.$$

$$A_{13'1} = \left\{\frac{-\rho U_0^2 S_4 b_1}{2\mathcal{M}_{zy}^{(3)}}\int_0^L \left(\frac{\partial c_\ell(\ell)}{\partial w}\right)_0^e \varphi_{xz}^{(1)} \varphi_{zy}^{(3)} d\ell\right\} s + \left\{\frac{\rho U_0^3 S_4 b_1}{2\mathcal{M}_{zy}^{(3)}}\int_0^L \left(\frac{\partial c_\ell(\ell)}{\partial w}\right)_0^e \sigma_{xz}^{(1)} \varphi_{zy}^{(3)} d\ell\right\}.$$

$$A_{13'2} = \left\{\frac{-\rho U_0^2 S_4 b_1}{2\mathcal{M}_{zy}^{(3)}}\int_0^L \left(\frac{\partial c_\ell(\ell)}{\partial w}\right)_0^e \varphi_{xz}^{(2)} \varphi_{zy}^{(3)} d\ell\right\} s + \left\{\frac{\rho U_0^3 S_4 b_1}{2\mathcal{M}_{zy}^{(3)}}\int_0^L \left(\frac{\partial c_\ell(\ell)}{\partial w}\right)_0^e \sigma_{xz}^{(2)} \varphi_{zy}^{(3)} d\ell\right\}.$$

$$A_{13'3} = \left\{\frac{-\rho U_0^2 S_4 b_1}{2\mathcal{M}_{zy}^{(3)}}\int_0^L \left(\frac{\partial c_\ell(\ell)}{\partial w}\right)_0^e \varphi_{xz}^{(3)} \varphi_{zy}^{(3)} d\ell\right\} s + \left\{\frac{\rho U_0^3 S_4 b_1}{2\mathcal{M}_{zy}^{(3)}}\int_0^L \left(\frac{\partial c_\ell(\ell)}{\partial w}\right)_0^e \sigma_{xz}^{(3)} \varphi_{zy}^{(3)} d\ell\right\}.$$

$$A_{13'4} = \left\{\frac{-\rho U_0^2 S_4 b_1}{2\mathcal{M}_{zy}^{(3)}}\int_0^L \left(\frac{\partial c_\ell(\ell)}{\partial w}\right)_0^e \varphi_{xz}^{(4)} \varphi_{zy}^{(3)} d\ell\right\} s + \left\{\frac{\rho U_0^3 S_4 b_1}{2\mathcal{M}_{zy}^{(3)}}\int_0^L \left(\frac{\partial c_\ell(\ell)}{\partial w}\right)_0^e \sigma_{xz}^{(4)} \varphi_{zy}^{(3)} d\ell\right\}.$$

$$A_{13'5} = \left\{\frac{-\rho U_0^2 S_4 b_1}{2\mathcal{M}_{zy}^{(3)}}\int_0^L \left(\frac{\partial c_\ell(\ell)}{\partial w}\right)_0^e \varphi_{xz}^{(5)} \varphi_{zy}^{(3)} d\ell\right\} s + \left\{\frac{\rho U_0^3 S_4 b_1}{2\mathcal{M}_{zy}^{(3)}}\int_0^L \left(\frac{\partial c_\ell(\ell)}{\partial w}\right)_0^e \sigma_{xz}^{(5)} \varphi_{zy}^{(3)} d\ell\right\}.$$

$$A_{13,6} = \left\{ \frac{-\rho U_0^2 S_4 b_1}{2\mathcal{M}_{zy}^{(3)}} \int_0^L \left(\frac{\partial c_\ell(\ell)}{\partial v}\right)_0^e \varphi_{xy}^{(1)} \varphi_{zy}^{(3)} d\ell \right\} s$$

$$+ \left\{ \frac{\rho U_0^3 S_4 b_1}{2\mathcal{M}_{zy}^{(3)}} \int_0^L \left(\frac{\partial c_\ell(\ell)}{\partial v}\right)_0^e \sigma_{xy}^{(1)} \varphi_{zy}^{(3)} d\ell \right\},$$

$$A_{13,7} = \left\{ \frac{-\rho U_0^2 S_4 b_1}{2\mathcal{M}_{zy}^{(3)}} \int_0^L \left(\frac{\partial c_\ell(\ell)}{\partial v}\right)_0^e \varphi_{xy}^{(2)} \varphi_{zy}^{(3)} d\ell \right\} s$$

$$+ \left\{ \frac{\rho U_0^3 S_4 b_1}{2\mathcal{M}_{zy}^{(3)}} \int_0^L \left(\frac{\partial c_\ell(\ell)}{\partial v}\right)_0^e \sigma_{xy}^{(2)} \varphi_{zy}^{(3)} d\ell \right\},$$

$$A_{13,8} = \left\{ \frac{-\rho U_0^2 S_4 b_1}{2\mathcal{M}_{zy}^{(3)}} \int_0^L \left(\frac{\partial c_\ell(\ell)}{\partial v}\right)_0^e \varphi_{xy}^{(3)} \varphi_{zy}^{(3)} d\ell \right\} s$$

$$+ \left\{ \frac{\rho U_0^3 S_4 b_1}{2\mathcal{M}_{zy}^{(3)}} \int_0^L \left(\frac{\partial c_\ell(\ell)}{\partial v}\right)_0^e \sigma_{xy}^{(3)} \varphi_{zy}^{(3)} d\ell \right\},$$

$$A_{13,9} = \left\{ \frac{-\rho U_0^2 S_4 b_1}{2\mathcal{M}_{zy}^{(3)}} \int_0^L \left(\frac{\partial c_\ell(\ell)}{\partial v}\right)_0^e \varphi_{xy}^{(4)} \varphi_{zy}^{(3)} d\ell \right\} s$$

$$+ \left\{ \frac{\rho U_0^3 S_4 b_1}{2\mathcal{M}_{zy}^{(3)}} \int_0^L \left(\frac{\partial c_\ell(\ell)}{\partial v}\right)_0^e \sigma_{xy}^{(4)} \varphi_{zy}^{(3)} d\ell \right\},$$

$$A_{13,10} = \left\{ \frac{-\rho U_0^2 S_4 b_1}{2\mathcal{M}_{zy}^{(3)}} \int_0^L \left(\frac{\partial c_\ell(\ell)}{\partial v}\right)_0^e \varphi_{xy}^{(5)} \varphi_{zy}^{(3)} d\ell \right\} s$$

$$+ \left\{ \frac{\rho U_0^3 S_4 b_1}{2\mathcal{M}_{zy}^{(3)}} \int_0^L \left(\frac{\partial c_\ell(\ell)}{\partial v}\right)_0^e \sigma_{xy}^{(5)} \varphi_{zy}^{(3)} d\ell \right\},$$

$$A_{13,11} = \left\{ \frac{-\rho U_0^2 S_4 b_1}{2\mathcal{M}_{zy}^{(3)}} \int_0^L \left(\frac{\partial c_\ell(\ell)}{\partial p}\right)_0^e \varphi_{zy}^{(1)} \varphi_{zy}^{(3)} d\ell \right\} s.$$

$$A_{13,12} = \left\{ \frac{-\rho U_0^2 S_4 b_1}{2\mathcal{M}_{zy}^{(3)}} \int_0^L \left(\frac{\partial c_\ell(\ell)}{\partial p}\right)_0^e \varphi_{zy}^{(2)} \varphi_{zy}^{(3)} d\ell \right\} s.$$

$$A_{13,13} = \{1\} s^2 + \left\{ 2\xi_{zy}^{(3)} \omega_{zy}^{(3)} - \frac{\rho U_0^2 S_4 b_1}{2\mathcal{M}_{zy}^{(3)}} \int_0^L \left(\frac{\partial c_\ell(\ell)}{\partial p}\right)_0^e \varphi_{zy}^{(3)2} d\ell \right\} s + \left\{ \omega_{zy}^{(3)2} \right\}.$$

$$A_{13'14} = \left\{ \frac{-\rho U_0^2 S_4 b_1}{2 \mathcal{M}_{zy}^{(3)}} \int_0^L \left(\frac{\partial C_\ell(\ell)}{\partial p} \right)_0^e \varphi_{zy}^{(4)} \varphi_{zy}^{(3)} d\ell \right\} s.$$

$$A_{13'15} = \left\{ \frac{-\rho U_0^2 S_4 b_1}{2 \mathcal{M}_{zy}^{(3)}} \int_0^L \left(\frac{\partial C_\ell(\ell)}{\partial p} \right)_0^e \varphi_{zy}^{(5)} \varphi_{zy}^{(3)} d\ell \right\} s.$$

$$A_{13'16} = \left\{ \frac{-\rho U_0^2 S_4 b_1}{2 \mathcal{M}_{zy}^{(3)}} \int_0^L \left[\left(\frac{\partial C_\ell(\ell)}{\partial u} \right)_0^e + \frac{2}{U_0} C_\ell(\ell)_0^e \right] \varphi_{zy}^{(3)} d\ell \right\}.$$

$$A_{13'17} = \left\{ \frac{-\rho U_0^2 S_4 b_1}{2 \mathcal{M}_{zy}^{(3)}} \int_0^L \left(\frac{\partial C_\ell(\ell)}{\partial v} \right)_0^e \varphi_{zy}^{(3)} d\ell \right\}.$$

$$A_{13'18} = \left\{ \frac{-\rho U_0^2 S_4 b_1}{2 \mathcal{M}_{zy}^{(3)}} \int_0^L \left(\frac{\partial C_\ell(\ell)}{\partial w} \right)_0^e \varphi_{zy}^{(3)} d\ell \right\}.$$

$$A_{13'19} = \left\{ \frac{-\rho U_0^2 S_4 b_1}{2 \mathcal{M}_{zy}^{(3)}} \int_0^L \left(\frac{\partial C_\ell(\ell)}{\partial p} \right)_0^e \varphi_{zy}^{(3)} d\ell \right\} s.$$

$$A_{13'20} = \left\{ \frac{-\rho U_0^2 S_4 b_1}{2 \mathcal{M}_{zy}^{(3)}} \int_0^L \left(\frac{\partial C_\ell(\ell)}{\partial q} \right)_0^e \varphi_{zy}^{(3)} d\ell \right\} s$$

$$A_{13'21} = \left\{ \frac{-\rho U_0^2 S_4 b_1}{2 \mathcal{M}_{zy}^{(3)}} \int_0^L \left(\frac{\partial C_\ell(\ell)}{\partial r} \right)_0^e \varphi_{zy}^{(3)} d\ell \right\} s$$

$A_{13'22} \cdots\cdots A_{13'40} = 0.$

$$A_{14'1} = \left\{ \frac{-\rho U_0^2 S_4 b_1}{2 \mathcal{M}_{zy}^{(4)}} \int_0^L \left(\frac{\partial C_\ell(\ell)}{\partial w} \right)_0^e \varphi_{xz}^{(1)} \varphi_{zy}^{(4)} d\ell \right\} s + \left\{ \frac{\rho U_0^3 S_4 b_1}{2 \mathcal{M}_{zy}^{(4)}} \int_0^L \left(\frac{\partial C_\ell(\ell)}{\partial w} \right)_0^e \sigma_{xz}^{(1)} \varphi_{zy}^{(4)} d\ell \right\},$$

$$A_{14'2} = \left\{ \frac{-\rho U_0^2 S_4 b_1}{2 \mathcal{M}_{zy}^{(4)}} \int_0^L \left(\frac{\partial C_\ell(\ell)}{\partial w} \right)_0^e \varphi_{xz}^{(2)} \varphi_{zy}^{(4)} d\ell \right\} s + \left\{ \frac{\rho U_0^3 S_4 b_1}{2 \mathcal{M}_{zy}^{(4)}} \int_0^L \left(\frac{\partial C_\ell(\ell)}{\partial w} \right)_0^e \sigma_{xz}^{(2)} \varphi_{zy}^{(4)} d\ell \right\},$$

$$A_{14'3} = \left\{ \frac{-\rho U_0^2 S_4 b_1}{2 \mathcal{M}_{zy}^{(4)}} \int_0^L \left(\frac{\partial C_\ell(\ell)}{\partial w} \right)_0^e \varphi_{xz}^{(3)} \varphi_{zy}^{(4)} d\ell \right\}_s$$

$$+ \left\{ \frac{\rho U_0^3 S_4 b_1}{2 \mathcal{M}_{zy}^{(4)}} \int_0^L \left(\frac{\partial C_\ell(\ell)}{\partial w} \right)_0^e \sigma_{xz}^{(3)} \varphi_{zy}^{(4)} d\ell \right\},$$

$$A_{14'4} = \left\{ \frac{-\rho U_0^2 S_4 b_1}{2 \mathcal{M}_{zy}^{(4)}} \int_0^L \left(\frac{\partial C_\ell(\ell)}{\partial w} \right)_0^e \varphi_{xz}^{(4)} \varphi_{zy}^{(4)} d\ell \right\}_s$$

$$+ \left\{ \frac{\rho U_0^3 S_4 b_1}{2 \mathcal{M}_{zy}^{(4)}} \int_0^L \left(\frac{\partial C_\ell(\ell)}{\partial w} \right)_0^e \sigma_{xz}^{(4)} \varphi_{zy}^{(4)} d\ell \right\},$$

$$A_{14'5} = \left\{ \frac{-\rho U_0^2 S_4 b_1}{2 \mathcal{M}_{zy}^{(4)}} \int_0^L \left(\frac{\partial C_\ell(\ell)}{\partial w} \right)_0^e \varphi_{xz}^{(5)} \varphi_{zy}^{(4)} d\ell \right\}_s$$

$$+ \left\{ \frac{\rho U_0^3 S_4 b_1}{2 \mathcal{M}_{zy}^{(4)}} \int_0^L \left(\frac{\partial C_\ell(\ell)}{\partial w} \right)_0^e \sigma_{xz}^{(5)} \varphi_{zy}^{(4)} d\ell \right\},$$

$$A_{14'6} = \left\{ \frac{-\rho U_0^2 S_4 b_1}{2 \mathcal{M}_{zy}^{(4)}} \int_0^L \left(\frac{\partial C_\ell(\ell)}{\partial v} \right)_0^e \varphi_{xy}^{(1)} \varphi_{zy}^{(4)} d\ell \right\}_s$$

$$+ \left\{ \frac{\rho U_0^3 S_4 b_1}{2 \mathcal{M}_{zy}^{(4)}} \int_0^L \left(\frac{\partial C_\ell(\ell)}{\partial v} \right)_0^e \sigma_{xy}^{(1)} \varphi_{zy}^{(4)} d\ell \right\},$$

$$A_{14'7} = \left\{ \frac{-\rho U_0^2 S_4 b_1}{2 \mathcal{M}_{zy}^{(4)}} \int_0^L \left(\frac{\partial C_\ell(\ell)}{\partial v} \right)_0^e \varphi_{xy}^{(2)} \varphi_{zy}^{(4)} d\ell \right\}_s$$

$$+ \left\{ \frac{\rho U_0^3 S_4 b_1}{2 \mathcal{M}_{zy}^{(4)}} \int_0^L \left(\frac{\partial C_\ell(\ell)}{\partial v} \right)_0^e \sigma_{xy}^{(2)} \varphi_{zy}^{(4)} d\ell \right\},$$

$$A_{14'8} = \left\{ \frac{-\rho U_0^2 S_4 b_1}{2 \mathcal{M}_{zy}^{(4)}} \int_0^L \left(\frac{\partial C_\ell(\ell)}{\partial v} \right)_0^e \varphi_{xy}^{(3)} \varphi_{zy}^{(4)} d\ell \right\}_s$$

$$+ \left\{ \frac{\rho U_0^3 S_4 b_1}{2 \mathcal{M}_{zy}^{(4)}} \int_0^L \left(\frac{\partial C_\ell(\ell)}{\partial v} \right)_0^e \sigma_{xy}^{(3)} \varphi_{zy}^{(4)} d\ell \right\},$$

$$A_{14'9} = \left\{ \frac{-\rho U_0^2 S_4 b_1}{2 \mathcal{M}_{zy}^{(4)}} \int_0^L \left(\frac{\partial C_\ell(\ell)}{\partial v} \right)_0^e \varphi_{xy}^{(4)} \varphi_{zy}^{(4)} d\ell \right\}_s$$

$$+ \left\{ \frac{\rho U_0^3 S_4 b_1}{2 \mathcal{M}_{zy}^{(4)}} \int_0^L \left(\frac{\partial C_\ell(\ell)}{\partial v} \right)_0^e \sigma_{xy}^{(4)} \varphi_{zy}^{(4)} d\ell \right\},$$

$$A_{14'10} = \left\{ \frac{-\rho U_0^2 S_4 b_1}{2\mathcal{M}_{zy}^{(4)}} \int_0^L \left(\frac{\partial c_\ell(\ell)}{\partial v} \right)_0^e \varphi_{xy}^{(5)} \varphi_{zy}^{(4)} d\ell \right\} s$$

$$+ \left\{ \frac{\rho U_0^3 S_4 b_1}{2\mathcal{M}_{zy}^{(4)}} \int_0^L \left(\frac{\partial c_\ell(\ell)}{\partial v} \right)_0^e \sigma_{xy}^{(5)} \varphi_{zy}^{(4)} d\ell \right\}.$$

$$A_{14'11} = \left\{ \frac{-\rho U_0^2 S_4 b_1}{2\mathcal{M}_{zy}^{(4)}} \int_0^L \left(\frac{\partial c_\ell(\ell)}{\partial p} \right)_0^e \varphi_{zy}^{(1)} \varphi_{zy}^{(4)} d\ell \right\} s.$$

$$A_{14'12} = \left\{ \frac{-\rho U_0^2 S_4 b_1}{2\mathcal{M}_{zy}^{(4)}} \int_0^L \left(\frac{\partial c_\ell(\ell)}{\partial p} \right)_0^e \varphi_{zy}^{(2)} \varphi_{zy}^{(4)} d\ell \right\} s.$$

$$A_{14'13} = \left\{ \frac{-\rho U_0^2 S_4 b_1}{2\mathcal{M}_{zy}^{(4)}} \int_0^L \left(\frac{\partial c_\ell(\ell)}{\partial p} \right)_0^e \varphi_{zy}^{(3)} \varphi_{zy}^{(4)} d\ell \right\} s.$$

$$A_{14'14} = \{1\} s^2 + \left\{ 2\xi_{zy}^{(4)} \omega_{zy}^{(4)} - \frac{\rho U_0^2 S_4 b_1}{2\mathcal{M}_{zy}^{(4)}} \int_0^L \left(\frac{\partial c_\ell(\ell)}{\partial p} \right)_0^e \varphi_{zy}^{(4)2} d\ell \right\} s + \left\{ \omega_{zy}^{(4)2} \right\}.$$

$$A_{14'15} = \left\{ \frac{-\rho U_0^2 S_4 b_1}{2\mathcal{M}_{zy}^{(4)}} \int_0^L \left(\frac{\partial c_\ell(\ell)}{\partial p} \right)_0^e \varphi_{zy}^{(5)} \varphi_{zy}^{(4)} d\ell \right\} s.$$

$$A_{14'16} = \left\{ \frac{-\rho U_0^2 S_4 b_1}{2\mathcal{M}_{zy}^{(4)}} \int_0^L \left[\left(\frac{\partial c_\ell(\ell)}{\partial u} \right)_0^e + \frac{2}{U_0} c_\ell(\ell)_0^e \right] \varphi_{zy}^{(4)} d\ell \right\}.$$

$$A_{14'17} = \left\{ \frac{-\rho U_0^2 S_4 b_1}{2\mathcal{M}_{zy}^{(4)}} \int_0^L \left(\frac{\partial c_\ell(\ell)}{\partial v} \right)_0^e \varphi_{zy}^{(4)} d\ell \right\}.$$

$$A_{14'18} = \left\{ \frac{-\rho U_0^2 S_4 b_1}{2\mathcal{M}_{zy}^{(4)}} \int_0^L \left(\frac{\partial c_\ell(\ell)}{\partial w} \right)_0^e \varphi_{zy}^{(4)} d\ell \right\}.$$

$$A_{14'19} = \left\{ \frac{-\rho U_0^2 S_4 b_1}{2\mathcal{M}_{zy}^{(4)}} \int_0^L \left(\frac{\partial c_\ell(\ell)}{\partial p} \right)_0^e \varphi_{zy}^{(4)} d\ell \right\} s$$

$$A_{14'20} = \left\{ \frac{-\rho U_0^2 S_4 b_1}{2\mathcal{M}_{zy}^{(4)}} \int_0^L \left(\frac{\partial c_\ell(\ell)}{\partial q} \right)_0^e \varphi_{zy}^{(4)} d\ell \right\} s$$

$$A_{14'21} = \left\{ \frac{-\rho U_0^2 S_4 b_1}{2\mathcal{M}_{zy}^{(4)}} \int_0^L \left(\frac{\partial c_\ell(\ell)}{\partial r} \right)_0^e \varphi_{zy}^{(4)} d\ell \right\} s$$

$A_{14'22} \cdots\cdots A_{14'40} = 0.$

$$A_{15'1} = \left\{ \frac{-\rho U_0^2 S_4 b_1}{2\mathcal{M}_{zy}^{(5)}} \int_0^L \left(\frac{\partial C_\ell(\ell)}{\partial w}\right)_0^e \varphi_{xz}^{(1)} \varphi_{zy}^{(5)} d\ell \right\}_s + \left\{ \frac{\rho U_0^3 S_4 b_1}{2\mathcal{M}_{zy}^{(5)}} \int_0^L \left(\frac{\partial C_\ell(\ell)}{\partial w}\right)_0^e \sigma_{xz}^{(1)} \varphi_{zy}^{(5)} d\ell \right\}.$$

$$A_{15'2} = \left\{ \frac{-\rho U_0^2 S_4 b_1}{2\mathcal{M}_{zy}^{(5)}} \int_0^L \left(\frac{\partial C_\ell(\ell)}{\partial w}\right)_0^e \varphi_{xz}^{(2)} \varphi_{zy}^{(5)} d\ell \right\}_s + \left\{ \frac{\rho U_0^3 S_4 b_1}{2\mathcal{M}_{zy}^{(5)}} \int_0^L \left(\frac{\partial C_\ell(\ell)}{\partial w}\right)_0^e \sigma_{xz}^{(2)} \varphi_{zy}^{(5)} d\ell \right\}.$$

$$A_{15'3} = \left\{ \frac{-\rho U_0^2 S_4 b_1}{2\mathcal{M}_{zy}^{(5)}} \int_0^L \left(\frac{\partial C_\ell(\ell)}{\partial w}\right)_0^e \varphi_{xz}^{(3)} \varphi_{zy}^{(5)} d\ell \right\}_s + \left\{ \frac{\rho U_0^3 S_4 b_1}{2\mathcal{M}_{zy}^{(5)}} \int_0^L \left(\frac{\partial C_\ell(\ell)}{\partial w}\right)_0^e \sigma_{xz}^{(3)} \varphi_{zy}^{(5)} d\ell \right\}.$$

$$A_{15'4} = \left\{ \frac{-\rho U_0^2 S_4 b_1}{2\mathcal{M}_{zy}^{(5)}} \int_0^L \left(\frac{\partial C_\ell(\ell)}{\partial w}\right)_0^e \varphi_{xz}^{(4)} \varphi_{zy}^{(5)} d\ell \right\}_s + \left\{ \frac{\rho U_0^3 S_4 b_1}{2\mathcal{M}_{zy}^{(5)}} \int_0^L \left(\frac{\partial C_\ell(\ell)}{\partial w}\right)_0^e \sigma_{xz}^{(4)} \varphi_{zy}^{(5)} d\ell \right\}.$$

$$A_{15'5} = \left\{ \frac{-\rho U_0^2 S_4 b_1}{2\mathcal{M}_{zy}^{(5)}} \int_0^L \left(\frac{\partial C_\ell(\ell)}{\partial w}\right)_0^e \varphi_{xz}^{(5)} \varphi_{zy}^{(5)} d\ell \right\}_s + \left\{ \frac{\rho U_0^3 S_4 b_1}{2\mathcal{M}_{zy}^{(5)}} \int_0^L \left(\frac{\partial C_\ell(\ell)}{\partial w}\right)_0^e \sigma_{xz}^{(5)} \varphi_{zy}^{(5)} d\ell \right\}.$$

$$A_{15'6} = \left\{ \frac{-\rho U_0^2 S_4 b_1}{2\mathcal{M}_{zy}^{(5)}} \int_0^L \left(\frac{\partial C_\ell(\ell)}{\partial v}\right)_0^e \varphi_{xy}^{(1)} \varphi_{zy}^{(5)} d\ell \right\}_s + \left\{ \frac{\rho U_0^3 S_4 b_1}{2\mathcal{M}_{zy}^{(5)}} \int_0^L \left(\frac{\partial C_\ell(\ell)}{\partial v}\right)_0^e \sigma_{xy}^{(1)} \varphi_{zy}^{(5)} d\ell \right\}.$$

$$A_{15'7} = \left\{ \frac{-\rho U_0^2 S_4 b_1}{2\mathcal{M}_{zy}^{(5)}} \int_0^L \left(\frac{\partial c_\ell(\ell)}{\partial v} \right)_0^e \varphi_{xy}^{(2)} \varphi_{zy}^{(5)} d\ell \right\} s$$

$$+ \left\{ \frac{\rho U_0^3 S_4 b_1}{2\mathcal{M}_{zy}^{(5)}} \int_0^L \left(\frac{\partial c_\ell(\ell)}{\partial v} \right)_0^e \sigma_{xy}^{(2)} \varphi_{zy}^{(5)} d\ell \right\},$$

$$A_{15'8} = \left\{ \frac{-\rho U_0^2 S_4 b_1}{2\mathcal{M}_{zy}^{(5)}} \int_0^L \left(\frac{\partial c_\ell(\ell)}{\partial v} \right)_0^e \varphi_{xy}^{(3)} \varphi_{zy}^{(5)} d\ell \right\} s$$

$$+ \left\{ \frac{\rho U_0^3 S_4 b_1}{2\mathcal{M}_{zy}^{(5)}} \int_0^L \left(\frac{\partial c_\ell(\ell)}{\partial v} \right)_0^e \sigma_{xy}^{(3)} \varphi_{zy}^{(5)} d\ell \right\},$$

$$A_{15'9} = \left\{ \frac{-\rho U_0^2 S_4 b_1}{2\mathcal{M}_{zy}^{(5)}} \int_0^L \left(\frac{\partial c_\ell(\ell)}{\partial v} \right)_0^e \varphi_{xy}^{(4)} \varphi_{zy}^{(5)} d\ell \right\} s$$

$$+ \left\{ \frac{\rho U_0^3 S_4 b_1}{2\mathcal{M}_{zy}^{(5)}} \int_0^L \left(\frac{\partial c_\ell(\ell)}{\partial v} \right)_0^e \sigma_{xy}^{(4)} \varphi_{zy}^{(5)} d\ell \right\},$$

$$A_{15'10} = \left\{ \frac{-\rho U_0^2 S_4 b_1}{2\mathcal{M}_{zy}^{(5)}} \int_0^L \left(\frac{\partial c_\ell(\ell)}{\partial v} \right)_0^e \varphi_{xy}^{(5)} \varphi_{zy}^{(5)} d\ell \right\} s$$

$$+ \left\{ \frac{\rho U_0^3 S_4 b_1}{2\mathcal{M}_{zy}^{(5)}} \int_0^L \left(\frac{\partial c_\ell(\ell)}{\partial v} \right)_0^e \sigma_{xy}^{(5)} \varphi_{zy}^{(5)} d\ell \right\},$$

$$A_{15'11} = \left\{ \frac{-\rho U_0^2 S_4 b_1}{2\mathcal{M}_{zy}^{(5)}} \int_0^L \left(\frac{\partial c_\ell(\ell)}{\partial p} \right)_0^e \varphi_{zy}^{(1)} \varphi_{zy}^{(5)} d\ell \right\} s.$$

$$A_{15'12} = \left\{ \frac{-\rho U_0^2 S_4 b_1}{2\mathcal{M}_{zy}^{(5)}} \int_0^L \left(\frac{\partial c_\ell(\ell)}{\partial p} \right)_0^e \varphi_{zy}^{(2)} \varphi_{zy}^{(5)} d\ell \right\} s.$$

$$A_{15'13} = \left\{ \frac{-\rho U_0^2 S_4 b_1}{2\mathcal{M}_{zy}^{(5)}} \int_0^L \left(\frac{\partial c_\ell(\ell)}{\partial p} \right)_0^e \varphi_{zy}^{(3)} \varphi_{zy}^{(5)} d\ell \right\} s.$$

$$A_{15'14} = \left\{ \frac{-\rho U_0^2 S_4 b_1}{2\mathcal{M}_{zy}^{(5)}} \int_0^L \left(\frac{\partial c_\ell(\ell)}{\partial p} \right)_0^e \varphi_{zy}^{(4)} \varphi_{zy}^{(5)} d\ell \right\} s.$$

$$A_{15'15} = \{1\} s^2 + \left\{ 2\xi_{zy}^{(5)} \omega_{zy}^{(5)} - \frac{\rho U_0^2 S_4 b_1}{2\mathcal{M}_{zy}^{(5)}} \int_0^L \left(\frac{\partial c_\ell(\ell)}{\partial p} \right)_0^e \varphi_{zy}^{(5)2} d\ell \right\} s + \left\{ \omega_{zy}^{(5)2} \right\}.$$

$$A_{15'16} = \left\{ \frac{-\rho U_0^2 S_4 b_1}{2\mathcal{M}_{zy}^{(5)}} \int_0^L \left[\left(\frac{\partial c_\ell(\ell)}{\partial u} \right)_0^e + \frac{2}{U_0} c_\ell(\ell)_0^e \right] \varphi_{zy}^{(5)} d\ell \right\}.$$

$$A_{15'17} = \left\{ \frac{-\rho U_0^2 S_4 b_1}{2 m_{zy}^{(5)}} \int_0^L \left(\frac{\partial c_\ell(\ell)}{\partial v} \right)_0^e \varphi_{zy}^{(5)} d\ell \right\}.$$

$$A_{15'18} = \left\{ \frac{-\rho U_0^2 S_4 b_1}{2 m_{zy}^{(5)}} \int_0^L \left(\frac{\partial c_\ell(\ell)}{\partial w} \right)_0^e \varphi_{zy}^{(5)} d\ell \right\}.$$

$$A_{15'19} = \left\{ \frac{-\rho U_0^2 S_4 b_1}{2 m_{zy}^{(5)}} \int_0^L \left(\frac{\partial c_\ell(\ell)}{\partial p} \right)_0^e \varphi_{zy}^{(5)} d\ell \right\}_s.$$

$$A_{15'20} = \left\{ \frac{-\rho U_0^2 S_4 b_1}{2 m_{zy}^{(5)}} \int_0^L \left(\frac{\partial c_\ell(\ell)}{\partial q} \right)_0^e \varphi_{zy}^{(5)} d\ell \right\}_s.$$

$$A_{15'21} = \left\{ \frac{-\rho U_0^2 S_4 b_1}{2 m_{zy}^{(5)}} \int_0^L \left(\frac{\partial c_\ell(\ell)}{\partial r} \right)_0^e \varphi_{zy}^{(5)} d\ell \right\}_s.$$

$$A_{15'22} \cdot \cdot \cdot \cdot \cdot A_{15'40} = 0.$$

$$A_{16'1} = \left\{ \frac{-\rho U_0 S_1}{2} \int_0^L \left[(c_L(\ell))_0^e - U_0 \left(\frac{\partial c_D(\ell)}{\partial w} \right)_0^e \right] \varphi_{xz}^{(1)} d\ell \right\}_s$$

$$+ \left\{ \frac{\rho U_0^2 S_1}{2} \int_0^L \left[(c_L(\ell))_0^e - U_0 \left(\frac{\partial c_D(\ell)}{\partial w} \right)_0^e \right] \sigma_{xz}^{(1)} d\ell \right\}.$$

$$A_{16'2} = \left\{ \frac{-\rho U_0 S_1}{2} \int_0^L \left[(c_L(\ell))_0^e - U_0 \left(\frac{\partial c_D(\ell)}{\partial w} \right)_0^e \right] \varphi_{xz}^{(2)} d\ell \right\}_s$$

$$+ \left\{ \frac{\rho U_0^2 S_1}{2} \int_0^L \left[(c_L(\ell))_0^e - U_0 \left(\frac{\partial c_D(\ell)}{\partial w} \right)_0^e \right] \sigma_{xz}^{(2)} d\ell \right\}.$$

$$A_{16'3} = \left\{ \frac{-\rho U_0 S_1}{2} \int_0^L \left[(c_L(\ell))_0^e - U_0 \left(\frac{\partial c_D(\ell)}{\partial w} \right)_0^e \right] \varphi_{xz}^{(3)} d\ell \right\}_s$$

$$+ \left\{ \frac{\rho U_0^2 S_1}{2} \int_0^L \left[(c_L(\ell))_0^e - U_0 \left(\frac{\partial c_D(\ell)}{\partial w} \right)_0^e \right] \sigma_{xz}^{(3)} d\ell \right\}.$$

$$A_{16'4} = \left\{ \frac{-\rho U_0 S_1}{2} \int_0^L \left[(c_L(\ell))_0^e - U_0 \left(\frac{\partial c_D(\ell)}{\partial w} \right)_0^e \right] \varphi_{xz}^{(4)} d\ell \right\}_s$$

$$+ \left\{ \frac{\rho U_0^2 S_1}{2} \int_0^L \left[(c_L(\ell))_0^e - U_0 \left(\frac{\partial c_D(\ell)}{\partial w} \right)_0^e \right] \sigma_{xz}^{(4)} d\ell \right\}$$

$$A_{16'5} = \left\{ \frac{-\rho U_0 S_1}{2} \int_0^L \left[(C_L(\ell))_0^e - U_0 \left(\frac{\partial C_D(\ell)}{\partial w} \right)_0^e \right] \varphi_{xz}^{(5)} \, d\ell \right\}_s$$

$$+ \left\{ \frac{\rho U_0^2 S_1}{2} \int_0^L \left[(C_L(\ell))_0^e - U_0 \left(\frac{\partial C_D(\ell)}{\partial w} \right)_0^e \right] \sigma_{xz}^{(5)} \, d\ell \right\}$$

$$A_{16'6} = \left\{ \frac{\rho U_0^2 S_1}{2} \int_0^L \left(\frac{\partial C_D(\ell)}{\partial v} \right)_0^e \varphi_{xy}^{(1)} \, d\ell \right\}_s$$

$$+ \left\{ \frac{-\rho U_0^3 S_1}{2} \int_0^L \left(\frac{\partial C_D(\ell)}{\partial v} \right)_0^e \sigma_{xy}^{(1)} \, d\ell \right\}.$$

$$A_{16'7} = \left\{ \frac{\rho U_0^2 S_1}{2} \int_0^L \left(\frac{\partial C_D(\ell)}{\partial v} \right)_0^e \varphi_{xy}^{(2)} \, d\ell \right\}_s$$

$$+ \left\{ \frac{-\rho U_0^3 S_1}{2} \int_0^L \left(\frac{\partial C_D(\ell)}{\partial v} \right)_0^e \sigma_{xy}^{(2)} \, d\ell \right\}.$$

$$A_{16'8} = \left\{ \frac{\rho U_0^2 S_1}{2} \int_0^L \left(\frac{\partial C_D(\ell)}{\partial v} \right)_0^e \varphi_{xy}^{(3)} \, d\ell \right\}_s$$

$$+ \left\{ \frac{-\rho U_0^3 S_1}{2} \int_0^L \left(\frac{\partial C_D(\ell)}{\partial v} \right)_0^e \sigma_{xy}^{(3)} \, d\ell \right\}.$$

$$A_{16'9} = \left\{ \frac{\rho U_0^2 S_1}{2} \int_0^L \left(\frac{\partial C_D(\ell)}{\partial v} \right)_0^e \varphi_{xy}^{(4)} \, d\ell \right\}_s$$

$$+ \left\{ \frac{-\rho U_0^3 S_1}{2} \int_0^L \left(\frac{\partial C_D(\ell)}{\partial v} \right)_0^e \sigma_{xy}^{(4)} \, d\ell \right\}.$$

$$A_{16'10} = \left\{ \frac{\rho U_0^2 S_1}{2} \int_0^L \left(\frac{\partial C_D(\ell)}{\partial v} \right)_0^e \varphi_{xy}^{(5)} \, d\ell \right\}_s$$

$$+ \left\{ \frac{-\rho U_0^3 S_1}{2} \int_0^L \left(\frac{\partial C_D(\ell)}{\partial v} \right)_0^e \sigma_{xy}^{(5)} \, d\ell \right\}.$$

$$A_{16'11} = \left\{ \frac{\rho U_0^2 S_1}{2} \int_0^L \left(\frac{\partial C_D(\ell)}{\partial p} \right)_0^e \varphi_{zy}^{(1)} \, d\ell \right\}_s.$$

$$A_{16'12} = \left\{ \frac{\rho U_0^2 S_1}{2} \int_0^L \left(\frac{\partial C_D(\ell)}{\partial p} \right)_0^e \varphi_{zy}^{(2)} \, d\ell \right\}_s.$$

$$A_{16'13} = \left\{ \frac{\rho U_0^2 S_1}{2} \int_0^L \left(\frac{\partial C_D(\ell)}{\partial p} \right)_0^e \varphi_{zy}^{(3)} \, d\ell \right\}_s.$$

$$A_{16',14} = \left\{ \frac{\rho U_0^2 S_1}{2} \int_0^L \left(\frac{\partial c_D(\ell)}{\partial p}\right)_0^e \varphi_{zy}^{(4)} d\ell \right\} s.$$

$$A_{16',15} = \left\{ \frac{\rho U_0^2 S_1}{2} \int_0^L \left(\frac{\partial c_D(\ell)}{\partial p}\right)_0^e \varphi_{zy}^{(5)} d\ell \right\} s$$

$$A_{16',16} = \left\{ M_t \right\} s + \left\{ \frac{\rho U_0^2 S_1}{2} \left[\left(\frac{\partial C_D}{\partial u}\right)_0^e + \frac{2}{U_0} (C_D)_0^e \right] \right\}.$$

$$A_{16',17} = \left\{ -MR_0 + \frac{\rho U_0^2 S_1}{2} \left(\frac{\partial C_D}{\partial v}\right)_0^e \right\}.$$

$$A_{16',18} = \left\{ MQ_0 - \frac{\rho U_0^2 S_1}{2} \left[\frac{1}{U_0} (C_L)_0^e - \left(\frac{\partial C_D}{\partial w}\right)_0^e \right] \right\}.$$

$$A_{16',19} = \left\{ \frac{\rho U_0^2 S_1}{2} \left(\frac{\partial C_D}{\partial p}\right)_0^e \right\} s.$$

$$A_{16',20} = \left\{ \frac{\rho U_0^2 S_1}{2} \left(\frac{\partial C_D}{\partial q}\right)_0^e \right\} s + \left\{ M_t\, g \cos \gamma_0 \cos \phi_0 \right\}.$$

$$A_{16',21} = \left\{ \frac{\rho U_0^2 S_1}{2} \left(\frac{\partial C_D}{\partial r}\right)_0^e - MV_0 \right\} s + \left\{ -M_t\, g \cos \gamma_0 \sin \phi_0 \right\}.$$

$$A_{16',22} \; \cdots\cdots \; A_{16',40} = 0.$$

$$A_{17',1} = \left\{ \frac{\rho U_0^2 S_2}{2} \int_0^L \left[\left(\frac{\partial c_y(\ell)}{\partial w}\right)_0^e + \left(\frac{\partial c_D(\ell)}{\partial w}\right)_0^e \beta_0 \right] \varphi_{xz}^{(1)} d\ell \right\} s.$$

$$+ \left\{ \frac{-\rho U_0^3 S_2}{2} \int_0^L \left[\left(\frac{\partial c_y(\ell)}{\partial w}\right)_0^e + \left(\frac{\partial c_D(\ell)}{\partial w}\right)_0^e \beta_0 \right] \sigma_{xz}^{(1)} d\ell \right\}.$$

$$A_{17',2} = \left\{ \frac{\rho U_0^2 S_2}{2} \int_0^L \left[\left(\frac{\partial c_y(\ell)}{\partial w}\right)_0^e + \left(\frac{\partial c_D(\ell)}{\partial w}\right)_0^e \beta_0 \right] \varphi_{xz}^{(2)} d\ell \right\} s.$$

$$+ \left\{ \frac{-\rho U_0^3 S_2}{2} \int_0^L \left[\left(\frac{\partial c_y(\ell)}{\partial w}\right)_0^e + \left(\frac{\partial c_D(\ell)}{\partial w}\right)_0^e \beta_0 \right] \sigma_{xz}^{(2)} d\ell \right\}.$$

$$A_{17',3} = \left\{ \frac{\rho U_0^2 S_2}{2} \int_0^L \left[\left(\frac{\partial c_y(\ell)}{\partial w}\right)_0^e + \left(\frac{\partial c_D(\ell)}{\partial w}\right)_0^e \beta_0 \right] \varphi_{xz}^{(3)} d\ell \right\} s$$

$$+ \left\{ \frac{-\rho U_0^3 S_2}{2} \int_0^L \left[\left(\frac{\partial c_y(\ell)}{\partial w}\right)_0^e + \left(\frac{\partial c_D(\ell)}{\partial w}\right)_0^e \beta_0 \right] \sigma_{xz}^{(3)} d\ell \right\}.$$

$$A_{17'4} = \left\{ \frac{\rho U_0^2 S_2}{2} \int_0^L \left[\left(\frac{\partial C_y(\ell)}{\partial w}\right)_0^e + \left(\frac{\partial C_D(\ell)}{\partial w}\right)_0^e \beta_0 \right] \varphi_{xz}^{(4)} \, d\ell \right\} s$$

$$+ \left\{ \frac{-\rho U_0^3 S_2}{2} \int_0^L \left[\left(\frac{\partial C_y(\ell)}{\partial w}\right)_0^e + \left(\frac{\partial C_D(\ell)}{\partial w}\right)_0^e \beta_0 \right] \sigma_{xz}^{(4)} \, d\ell \right\}.$$

$$A_{17'5} = \left\{ \frac{\rho U_0^2 S_2}{2} \int_0^L \left[\left(\frac{\partial C_y(\ell)}{\partial w}\right)_0^e + \left(\frac{\partial C_D(\ell)}{\partial w}\right)_0^e \beta_0 \right] \varphi_{xz}^{(5)} \, d\ell \right\} s$$

$$+ \left\{ \frac{-\rho U_0^3 S_2}{2} \int_0^L \left[\left(\frac{\partial C_y(\ell)}{\partial w}\right)_0^e + \left(\frac{\partial C_D(\ell)}{\partial w}\right)_0^e \beta_0 \right] \sigma_{xz}^{(5)} \, d\ell \right\}.$$

$$A_{17'6} = \left\{ \frac{\rho U_0^2 S_2}{2} \int_0^L \left[\left(\frac{\partial C_y(\ell)}{\partial v}\right)_0^e + \left(\frac{\partial C_D(\ell)}{\partial v}\right)_0^e \beta_0 + \frac{1}{U_0} (C_D(\ell))_0^e \right] \varphi_{xy}^{(1)} d\ell \right\} s$$

$$+ \left\{ \varphi_{xy}^{(1)}(eG) - (\ell_R) \sigma_{xy}^{(1)}(eG) \right\} M_R s^2 + \left\{ \frac{-\rho U_0^3 S_2}{2} \int_0^L \left[\left(\frac{\partial C_y(\ell)}{\partial v}\right)_0^e + \left(\frac{\partial C_D(\ell)}{\partial v}\right)_0^e \beta_0 \right.\right.$$

$$\left.\left. + \frac{1}{U_0} (C_D(\ell))_0^e \right] \sigma_{xy}^{(1)} d\ell - \sigma_{xy}^{(1)}(eG) \, (T_s + T_c) \right\}.$$

$$A_{17'7} = \left\{ \frac{\rho U_0^2 S_2}{2} \int_0^L \left[\left(\frac{\partial C_y(\ell)}{\partial v}\right)_0^e + \left(\frac{\partial C_D(\ell)}{\partial v}\right)_0^e \beta_0 + \frac{1}{U_0} (C_D(\ell))_0^e \right] \varphi_{xy}^{(2)} d\ell \right\} s$$

$$+ \left\{ M_R \left[\varphi_{xy}^{(2)}(eG) - \ell_R \, \sigma_{xy}^{(2)}(eG) \right] \right\} s^2 + \left\{ \frac{-\rho U_0^3 S_2}{2} \int_0^L \left[\left(\frac{\partial C_y(\ell)}{\partial v}\right)_0^e + \left(\frac{\partial C_D(\ell)}{\partial v}\right)_0^e \beta_0 \right.\right.$$

$$\left.\left. + \frac{1}{U_0} (C_D(\ell))_0^e \right] \sigma_{xy}^{(2)} d\ell - (T_s + T_c) \, \sigma_{xy}^{(2)}(eG) \right\}.$$

$$A_{17'8} = \left\{ \frac{\rho U_0^2 S_2}{2} \int_0^L \left[\left(\frac{\partial C_y(\ell)}{\partial v}\right)_0^e + \left(\frac{\partial C_D(\ell)}{\partial v}\right)_0^e \beta_0 + \frac{1}{U_0} (C_D(\ell))_0^e \right] \varphi_{xy}^{(3)} d\ell \right\} s$$

$$+ \left\{ M_R \left[\varphi_{xy}^{(3)}(eG) - \ell_R \, \sigma_{xy}^{(3)}(eG) \right] \right\} s^2 + \left\{ \frac{-\rho U_0^3 S_2}{2} \int_0^L \left[\left(\frac{\partial C_y(\ell)}{\partial v}\right)_0^e + \left(\frac{\partial C_D(\ell)}{\partial v}\right)_0^e \beta_0 \right.\right.$$

$$\left.\left. + \frac{1}{U_0} (C_D(\ell))_0^e \right] \sigma_{xy}^{(3)} d\ell - (T_s + T_c) \, \sigma_{xy}^{(3)}(eG) \right\}.$$

$$_{7'9} = \left\{ \frac{\rho U_0^2 S_2}{2} \int_0^L \left[\left(\frac{\partial C_y(\ell)}{\partial v}\right)_0^e + \left(\frac{\partial C_D(\ell)}{\partial v}\right)_0^e \beta_0 + \frac{1}{U_0} (C_D(\ell))_0^e \right] \varphi_{xy}^{(4)} d\ell \right\} s$$

$$+ \left\{ M_R \left[\varphi_{xy}^{(4)}(eG) - \ell_R \sigma_{xy}^{(4)}(eG) \right] \right\} s^2 + \left\{ \frac{-\rho U_0^3 S_2}{2} \int_0^L \left[\left(\frac{\partial C_y(\ell)}{\partial v}\right)_0^e + \left(\frac{\partial C_D(\ell)}{\partial v}\right)_0^e \beta_0 \right.\right.$$

$$\left.\left. + \frac{1}{U_0} (C_D(\ell))_0^e \right] \sigma_{xy}^{(4)} d\ell - (T_s + T_c) \sigma_{xy}^{(4)}(eG) \right\}.$$

$$A_{17'10} = \left\{ \frac{\rho U_0^2 S_2}{2} \int_0^L \left[\left(\frac{\partial C_y(\ell)}{\partial v}\right)_0^e + \left(\frac{\partial C_D(\ell)}{\partial v}\right)_0^e \beta_0 + \frac{1}{U_0} (C_D(\ell))_0^e \right] \varphi_{xy}^{(5)} d\ell \right\} s$$

$$+ \left\{ M_R \left[\varphi_{xy}^{(5)}(eG) - \ell_R \sigma_{xy}^{(5)}(eG) \right] \right\} s^2 + \left\{ \frac{-\rho U_0^3 S_2}{2} \int_0^L \left[\left(\frac{\partial C_y(\ell)}{\partial v}\right)_0^e + \left(\frac{\partial C_D(\ell)}{\partial v}\right)_0^e \beta_0 \right.\right.$$

$$\left.\left. + \frac{1}{U_0} (C_D(\ell))_0^e \right] \sigma_{xy}^{(5)} d\ell - (T_s + T_c) \sigma_{xy}^{(5)}(eG) \right\}.$$

$$A_{17'11} = \left\{ \frac{\rho U_0^2 S_2}{2} \int_0^L \left[\left(\frac{\partial C_y(\ell)}{\partial p}\right)_0^e + \left(\frac{\partial C_D(\ell)}{\partial p}\right)_0^e \beta_0 \right] \varphi_{zy}^{(1)} d\ell \right\} s.$$

$$A_{17'12} \quad \left\{ \frac{\rho U_0^2 S_2}{2} \int_0^L \left[\left(\frac{\partial C_y(\ell)}{\partial p}\right)_0^e + \left(\frac{\partial C_D(\ell)}{\partial p}\right)_0^e \beta_0 \right] \varphi_{zy}^{(2)} d\ell \right\} s.$$

$$A_{17'13} = \left\{ \frac{\rho U_0^2 S_2}{2} \int_0^L \left[\left(\frac{\partial C_y(\ell)}{\partial p}\right)_0^e + \left(\frac{\partial C_D(\ell)}{\partial p}\right)_0^e \beta_0 \right] \varphi_{zy}^{(3)} d\ell \right\} s.$$

$$A_{17'14} = \left\{ \frac{\rho U_0^2 S_2}{2} \int_0^L \left[\left(\frac{\partial C_y(\ell)}{\partial p}\right)_0^e + \left(\frac{\partial C_D(\ell)}{\partial p}\right)_0^e \beta_0 \right] \varphi_{zy}^{(4)} d\ell \right\} s.$$

$$A_{17'15} = \left\{ \frac{\rho U_0^2 S_2}{2} \int_0^L \left[\left(\frac{\partial C_y(\ell)}{\partial p}\right)_0^e + \left(\frac{\partial C_D(\ell)}{\partial p}\right)_0^e \beta_0 \right] \varphi_{zy}^{(5)} d\ell \right\} s.$$

$$A_{17'16} = \left\{ MR_0 + \frac{\rho U_0^2 S_2}{2} \left[\left(\frac{\partial C_y}{\partial u}\right)_0^e + \left(\frac{\partial C_D}{\partial u}\right)_0^e \beta_0 + \frac{2}{U_0} (C_y)_0^e + \frac{2}{U_0} \beta_0 (C_D)_0^e \right] \right\}$$

$$A_{17'17} = \left\{ M + M_R + \frac{\rho U_0^2 S_2}{2} \left[\left(\frac{\partial C_y}{\partial \dot{v}}\right)_0^e + \left(\frac{\partial C_D}{\partial \dot{v}}\right)_0^e \beta_0 \right] \right\} s$$

$$+ \left\{ \frac{\rho U_0^2 S_2}{2} \left[\left(\frac{\partial C_y}{\partial v}\right)_0^e + \left(\frac{\partial C_D}{\partial v}\right)_0^e \beta_0 + \frac{1}{U_0} (C_D)_0^e \right] \right\}.$$

$$A_{17'18} = \left\{ -MP_0 + \frac{\rho U_0^2 S_2}{2} \left[\left(\frac{\partial C_y}{\partial w}\right)_0^e + \left(\frac{\partial C_D}{\partial w}\right)_0^e \beta_0 \right] \right\}$$

$$A_{17,19} = \left\{-M_t\, g \cos\gamma_0 \cos\phi_0\right\} + \left\{\frac{\rho U_0^2 S_2}{2}\left[\left(\frac{\partial C_y}{\partial p}\right)_0^e + \left(\frac{\partial C_D}{\partial p}\right)_0^e \beta_0\right]\right\} s.$$

$$A_{17,20} = \left\{\frac{\rho U_0^2 S_2}{2}\left[\left(\frac{\partial C_y}{\partial q}\right)_0^e + \left(\frac{\partial C_D}{\partial q}\right)_0^e \beta_0\right]\right\} s.$$

$$A_{17,21} = \left\{-M_R (L - \ell_{eG})\right\} s^2 + \left\{M U_0 + \frac{\rho U_0^2 S_2}{2}\left[\left(\frac{\partial C_y}{\partial r}\right)_0^e + \left(\frac{\partial C_D}{\partial r}\right)_0^e \beta_0\right]\right\} s$$

$$+\left\{-M_t\, g \sin\gamma_0\right\}.$$

$A_{17,22} \cdots A_{17,25} = 0.$

$A_{17,26} = K_1$

$A_{17,27} = K_2$

$A_{17,28} = K_3$

$A_{17,29} = K_4$

$A_{17,30} = 0.$

$A_{17,31} = \left\{M_R\, \ell_R\right\} s^2 + \left\{T_c\right\}.$

$A_{17,32} \cdots A_{17,40} = 0.$

$$A_{18,1} = \left\{M_R\left[\varphi_{xz}^{(1)}(eG) - \ell_R\, \sigma_{xz}^{(1)}(eG)\right]\right\} s^2 + \left\{\frac{\rho U_0^2 S_3}{2}\int_0^L\left[\left(\frac{\partial C_L(\ell)}{\partial w}\right)_0^e + \frac{1}{U_0} C_D(\ell)_0^e\right]\varphi_{xz}^{(1)} d\ell\right\} s$$

$$+\left\{-(T_s + T_c)\,\sigma_{xz}^{(1)}(eG) - \frac{\rho U_0^3 S_3}{2}\int_0^L\left[\left(\frac{\partial C_L(\ell)}{\partial w}\right)_0^e + \frac{1}{U_0} C_D(\ell)_0^e\right]\sigma_{xz}^{(1)}\, d\ell\right\}.$$

$$A_{18,2} = \left\{M_R\left[\varphi_{xz}^{(2)}(eG) - \ell_R\, \sigma_{xz}^{(2)}(eG)\right]\right\} s^2 + \left\{\frac{\rho U_0^2 S_3}{2}\int_0^L\left[\left(\frac{\partial C_L(\ell)}{\partial w}\right)_0^e + \frac{1}{U_0} C_D(\ell)_0^e\right]\varphi_{xz}^{(2)} d\ell\right\} s$$

$$+\left\{-(T_s + T_c)\,\sigma_{xz}^{(2)}(eG) - \frac{\rho U_0^3 S_3}{2}\int_0^L\left[\left(\frac{\partial C_L(\ell)}{\partial w}\right)_0^e + \frac{1}{U_0} C_D(\ell)_0^e\right]\sigma_{xz}^{(2)}\, d\ell\right\}$$

$$A_{18,3} = \left\{M_R\left[\varphi_{xz}^{(3)}(eG) - \ell_R\, \sigma_{xz}^{(3)}(eG)\right]\right\} s^2 + \left\{\frac{\rho U_0^2 S_3}{2}\int_0^L\left[\left(\frac{\partial C_L(\ell)}{\partial w}\right)_0^e + \frac{1}{U_0} C_D(\ell)_0^e\right]\varphi_{xz}^{(3)} d\ell\right\} s$$

$$+\left\{-(T_s + T_c)\,\sigma_{xz}^{(3)}(eG) - \frac{\rho U_0^3 S_3}{2}\int_0^L\left[\left(\frac{\partial C_L(\ell)}{\partial w}\right)_0^e + \frac{1}{U_0} C_D(\ell)_0^e\right]\sigma_{xz}^{(3)}\, d\ell\right\}.$$

$$A_{18,4} = \left\{ M_R \left[\varphi_{xz}^{(4)}(eG) - \ell_R \, \sigma_{xz}^{(4)}(eG) \right] \right\} s^2 + \left\{ \frac{\rho U_0^2 S_3}{2} \int_0^L \left[\left(\frac{\partial C_L(\ell)}{\partial w} \right)_0^e + \frac{1}{U_0} C_D(\ell)_0^e \right] \varphi_{xz}^{(4)} d\ell \right\} s$$

$$+ \left\{ -(T_s + T_c) \, \sigma_{xz}^{(4)}(eG) - \frac{\rho U_0^3 S_3}{2} \int_0^L \left[\left(\frac{\partial C_L(\ell)}{\partial w} \right)_0^e + \frac{1}{U_0} C_D(\ell)_0^e \right] \sigma_{xz}^{(4)} d\ell \right\}$$

$$A_{18,5} = \left\{ M_R \left[\varphi_{xz}^{(5)}(eG) - \ell_R \, \sigma_{xz}^{(5)}(eG) \right] \right\} s^2 + \left\{ \frac{\rho U_0^2 S_3}{2} \int_0^L \left[\left(\frac{\partial C_L(\ell)}{\partial w} \right)_0^e + \frac{1}{U_0} C_D(\ell)_0^e \right] \varphi_{xz}^{(5)} d\ell \right\} s$$

$$+ \left\{ -(T_s + T_c) \, \sigma_{xz}^{(5)}(eG) - \frac{\rho U_0^3 S_3}{2} \int_0^L \left[\left(\frac{\partial C_L(\ell)}{\partial w} \right)_0^e + \frac{1}{U_0} C_D(\ell)_0^e \right] \sigma_{xz}^{(5)} d\ell \right\}.$$

$$A_{18,6} = \left\{ \frac{\rho U_0^2 S_3}{2} \int_0^L \left(\frac{\partial C_L(\ell)}{\partial v} \right)_0^e \varphi_{xy}^{(1)} d\ell \right\} s + \left\{ \frac{-\rho U_0^3 S_3}{2} \int_0^L \left(\frac{\partial C_L(\ell)}{\partial v} \right)_0^e \sigma_{xy}^{(1)} d\ell \right\}.$$

$$A_{18,7} = \left\{ \frac{\rho U_0^2 S_3}{2} \int_0^L \left(\frac{\partial C_L(\ell)}{\partial v} \right)_0^e \varphi_{xy}^{(2)} d\ell \right\} s + \left\{ \frac{-\rho U_0^3 S_3}{2} \int_0^L \left[\left(\frac{\partial C_L(\ell)}{\partial v} \right)_0^e \sigma_{xy}^{(2)} d\ell \right\}.$$

$$A_{18,8} = \left\{ \frac{\rho U_0^2 S_3}{2} \int_0^L \left(\frac{\partial C_L(\ell)}{\partial v} \right)_0^e \varphi_{xy}^{(3)} d\ell \right\} s + \left\{ \frac{-\rho U_0^3 S_3}{2} \int_0^L \left(\frac{\partial C_L(\ell)}{\partial v} \right)_0^e \sigma_{xy}^{(3)} d\ell \right\}.$$

$$A_{18,9} = \left\{ \frac{\rho U_0^2 S_3}{2} \int_0^L \left(\frac{\partial C_L(\ell)}{\partial v} \right)_0^e \varphi_{xy}^{(4)} d\ell \right\} s + \left\{ \frac{-\rho U_0^3 S_3}{2} \int_0^L \left[\left(\frac{\partial C_L(\ell)}{\partial v} \right)_0^e \sigma_{xy}^{(4)} d\ell \right\}$$

$$A_{18,10} = \left\{ \frac{\rho U_0^2 S_3}{2} \int_0^L \left(\frac{\partial C_L(\ell)}{\partial v} \right)_0^e \varphi_{xy}^{(5)} d\ell \right\} s + \left\{ \frac{-\rho U_0^3 S_3}{2} \int_0^L \left(\frac{\partial C_L(\ell)}{\partial v} \right)_0^e \sigma_{xy}^{(5)} d\ell \right\}.$$

$$A_{18,11} = \left\{ \frac{\rho U_0^2 S_3}{2} \int_0^L \left(\frac{\partial C_L(\ell)}{\partial p} \right)_0^e \varphi_{zy}^{(1)} d\ell \right\} s.$$

$$A_{18,12} = \left\{ \frac{\rho U_0^2 S_3}{2} \int_0^L \left(\frac{\partial C_L(\ell)}{\partial p} \right)_0^e \varphi_{zy}^{(2)} d\ell \right\} s.$$

$$A_{18,13} = \left\{ \frac{\rho U_0^2 S_3}{2} \int_0^L \left(\frac{\partial C_L(\ell)}{\partial p} \right)_0^e \varphi_{zy}^{(3)} d\ell \right\} s.$$

$$A_{18,14} = \left\{ \frac{\rho U_0^2 S_3}{2} \int_0^L \left(\frac{\partial C_L(\ell)}{\partial p} \right)_0^e \varphi_{zy}^{(4)} d\ell \right\} s.$$

$$A_{18,15} = \left\{ \frac{\rho U_0^2 S_3}{2} \int_0^L \left(\frac{\partial C_L(\ell)}{\partial p} \right)_0^e \varphi_{zy}^{(5)} d\ell \right\} s.$$

$$A_{18'16} = \left\{ -MQ_0 + \frac{\rho U_0^2 S_3}{2} \left[\left(\frac{\partial C_L}{\partial u} \right)_0^e + \frac{2}{U_0} (C_L)_0^e \right] \right\}.$$

$$A_{18'17} = \left\{ MP_0 + \frac{\rho U_0^2 S_3}{2} \left(\frac{\partial C_L}{\partial v} \right)_0^e \right\}.$$

$$A_{18'18} = \left\{ M + M_R + \frac{\rho U_0^2 S_3}{2} \left(\frac{\partial C_L}{\partial \dot{w}} \right)_0^e \right\} s + \left\{ \frac{\rho U_0^2 S_3}{2} \left[\left(\frac{\partial C_L}{\partial w} \right)_0^e + \frac{1}{U_0} (C_D)_0^e \right] \right\}.$$

$$A_{18'19} = \left\{ MV_0 + \frac{\rho U_0^2 S_3}{2} \left(\frac{\partial C_L}{\partial p} \right)_0^e \right\} s + \left\{ M_t g \cos \gamma_0 \sin \phi_0 \right\}.$$

$$A_{18'20} = \left\{ M_R (L - \ell_{CG}) \right\} s^2 + \left\{ -MU_0 + \frac{\rho U_0^2 S_3}{2} \left(\frac{\partial C_L}{\partial q} \right)_0^e \right\} s + \left\{ M_t g \sin \gamma_0 \right\}.$$

$$A_{18'21} = \left\{ \frac{\rho U_0^2 S_3}{2} \left(\frac{\partial C_L}{\partial r} \right)_0^e \right\} s.$$

$$A_{18'22} = K_1.$$

$$A_{18'23} = K_2.$$

$$A_{18'24} = K_3.$$

$$A_{18'25} = K_4.$$

$$A_{18'26} \cdots A_{18'29} = 0.$$

$$A_{18'30} = \left\{ M_R \ell_R \right\} s^2 + \left\{ T_c \right\}.$$

$$A_{18'31} \cdots A_{18'40} = 0.$$

$$A_{19'1} = \left\{ \frac{-\rho U_0^2 S_4 b_1}{2} \int_0^L \left(\frac{\partial c_\ell(\ell)}{\partial w} \right)_0^e \varphi_{xz}^{(1)} d\ell \right\} s + \left\{ \frac{\rho U_0^3 S_4 b_1}{2} \int_0^L \left[\left(\frac{\partial c_\ell(\ell)}{\partial w} \right)_0^e \sigma_{xz}^{(1)} d\ell \right\}$$

$$A_{19'2} = \left\{ \frac{-\rho U_0^2 S_4 b_1}{2} \int_0^L \left(\frac{\partial c_\ell(\ell)}{\partial w} \right)_0^e \varphi_{xz}^{(2)} d\ell \right\} s + \left\{ \frac{\rho U_0^3 S_4 b_1}{2} \int_0^L \left[\left(\frac{\partial c_\ell(\ell)}{\partial w} \right)_0^e \sigma_{xz}^{(2)} d\ell \right\}$$

$$A_{19'3} = \left\{ \frac{-\rho U_0^2 S_4 b_1}{2} \int_0^L \left(\frac{\partial c_\ell(\ell)}{\partial w} \right)_0^e \varphi_{xz}^{(3)} d\ell \right\} s + \left\{ \frac{\rho U_0^3 S_4 b_1}{2} \int_0^L \left(\frac{\partial c_\ell(\ell)}{\partial w} \right)_0^e \sigma_{xz}^{(3)} d\ell \right\}$$

$$A_{19'4} = \left\{ \frac{-\rho U_0^2 S_4 b_1}{2} \int_0^L \left(\frac{\partial c_\ell(\ell)}{\partial w} \right)_0^e \varphi_{xz}^{(4)} d\ell \right\} s + \left\{ \frac{\rho U_0^3 S_4 b_1}{2} \int_0^L \left[\left(\frac{\partial c_\ell(\ell)}{\partial w} \right)_0^e \sigma_{xz}^{(4)} d\ell \right\}$$

$$A_{19'5} = \left\{ \frac{-\rho U_0^2 S_4 b_1}{2} \int_0^L \left(\frac{\partial c_\ell(\ell)}{\partial w} \right)_0^e \varphi_{xz}^{(5)} d\ell \right\} s + \left\{ \frac{\rho U_0^3 S_4 b_1}{2} \int_0^L \left[\left(\frac{\partial c_\ell(\ell)}{\partial w} \right)_0^e \sigma_{xz}^{(5)} d\ell \right\}$$

$$A_{19'6} = \left\{ \frac{-\rho U_0^2 S_4 b_1}{2} \int_0^L \left(\frac{\partial c_\ell(\ell)}{\partial v} \right)_0^e \varphi_{xy}^{(1)} d\ell \right\} s + \left\{ \frac{\rho U_0^3 S_4 b_1}{2} \int_0^L \left[\left(\frac{\partial c_\ell(\ell)}{\partial v} \right)_0^e \sigma_{xy}^{(1)} d\ell \right\}$$

$$A_{19',7} = \left\{ \frac{-\rho U_0^2 S_4 b_1}{2} \int_0^L \left(\frac{\partial c_\ell(\ell)}{\partial v} \right)_0^e \varphi_{xy}^{(2)} d\ell \right\} s + \left\{ \frac{\rho U_0^3 S_4 b_1}{2} \int_0^L \left[\left(\frac{\partial c_\ell(\ell)}{\partial v} \right)_0^e \sigma_{xy}^{(2)} d\ell \right\}$$

$$A_{19',8} = \left\{ \frac{-\rho U_0^2 S_4 b_1}{2} \int_0^L \left(\frac{\partial c_\ell(\ell)}{\partial v} \right)_0^e \varphi_{xy}^{(3)} d\ell \right\} s + \left\{ \frac{\rho U_0^3 S_4 b_1}{2} \int_0^L \left[\left(\frac{\partial c_\ell(\ell)}{\partial v} \right)_0^e \sigma_{xy}^{(3)} d\ell \right\}$$

$$A_{19',9} = \left\{ \frac{-\rho U_0^2 S_4 b_1}{2} \int_0^L \left(\frac{\partial c_\ell(\ell)}{\partial v} \right)_0^e \varphi_{xy}^{(4)} d\ell \right\} s + \left\{ \frac{\rho U_0^3 S_4 b_1}{2} \int_0^L \left[\left(\frac{\partial c_\ell(\ell)}{\partial v} \right)_0^e \sigma_{xy}^{(4)} d\ell \right\}$$

$$A_{19',10} = \left\{ \frac{-\rho U_0^2 S_4 b_1}{2} \int_0^L \left(\frac{\partial c_\ell(\ell)}{\partial v} \right)_0^e \varphi_{xy}^{(5)} d\ell \right\} s + \left\{ \frac{\rho U_0^3 S_4 b_1}{2} \int_0^L \left[\left(\frac{\partial c_\ell(\ell)}{\partial v} \right)_0^e \sigma_{xy}^{(5)} d\ell \right\}$$

$$A_{19',11} = \left\{ \frac{-\rho U_0^2 S_4 b_1}{2} \int_0^L \left(\frac{\partial c_\ell(\ell)}{\partial p} \right)_0^e \varphi_{zy}^{(1)} d\ell \right\} s.$$

$$A_{19',12} = \left\{ \frac{-\rho U_0^2 S_4 b_1}{2} \int_0^L \left(\frac{\partial c_\ell(\ell)}{\partial p} \right)_0^e \varphi_{zy}^{(2)} d\ell \right\} s.$$

$$A_{19',13} = \left\{ \frac{-\rho U_0^2 S_4 b_1}{2} \int_0^L \left(\frac{\partial c_\ell(\ell)}{\partial p} \right)_0^e \varphi_{zy}^{(3)} d\ell \right\} s.$$

$$A_{19',14} = \left\{ \frac{-\rho U_0^2 S_4 b_1}{2} \int_0^L \left(\frac{\partial c_\ell(\ell)}{\partial p} \right)_0^e \varphi_{zy}^{(4)} d\ell \right\} s.$$

$$A_{19',15} = \left\{ \frac{-\rho U_0^2 S_4 b_1}{2} \int_0^L \left(\frac{\partial c_\ell(\ell)}{\partial p} \right)_0^e \varphi_{zy}^{(5)} d\ell \right\} s.$$

$$A_{19',16} = \left\{ \frac{-\rho U_0^2 S_4 b_1}{2} \left[\left(\frac{\partial c_\ell}{\partial u} \right)_0^e + \frac{2}{U_0} (c_\ell)_0^e \right] \right\}.$$

$$A_{19',17} = \left\{ \frac{-\rho U_0^2 S_4 b_1}{2} \left(\frac{\partial c_\ell}{\partial v} \right)_0^e \right\}.$$

$$A_{19',18} = \left\{ \frac{-\rho U_0^2 S_4 b_1}{2} \left(\frac{\partial c_\ell}{\partial w} \right)_0^e \right\}.$$

$$A_{19',19} = \left\{ I_{xx} \right\} s^2 + \left\{ -I_{xz} Q_0 - \frac{\rho U_0^2 S_4 b_1}{2} \left(\frac{\partial c_\ell}{\partial p} \right)_0^e \right\} s.$$

$$A_{19',20} = \left\{ (I_{zz} - I_{yy}) R_0 - I_{xz} P_0 - \frac{\rho U_0^2 S_4 b_1}{2} \left(\frac{\partial c_\ell}{\partial q} \right)_0^e \right\} s.$$

$$A_{19',21} = \left\{ -I_{xz} \right\} s^2 + \left\{ (I_{zz} - I_{yy}) Q_0 - \frac{\rho U_0^2 S_4 b_1}{2} \left(\frac{\partial c_\ell}{\partial r} \right)_0^e \right\} s.$$

$$A_{19',22} \cdots\cdots A_{19',40} = 0.$$

$$A_{20'1} = \left\{ M_R (L - \ell_{CG}) \left[\varphi_{xz}^{(1)}(eG) - \ell_R \, \sigma_{xz}^{(1)}(eG) \right] \right\} s^2$$

$$+ \left\{ - \frac{\rho U_0^2 S_5 b_2}{2} \left[\int_0^L \left(\frac{\partial C_m(\ell)}{\partial w} \right)_0^e \varphi_{xz}^{(1)} d\ell - \int_0^L \left(\frac{\partial C_m(\ell)}{\partial \dot{w}} \right)_0^e U_0 \, \sigma_{xz}^{(1)} d\ell \right] \right\} s$$

$$+ \left\{ -\ell_e (T_s + T_c) \sigma_{xz}^{(1)}(eG) - (T_s + T_c) \varphi_{xz}^{(1)}(eG) + \frac{\rho U_0^3 S_5 b_2}{2} \int_0^L \left(\frac{\partial C_m(\ell)}{\partial w} \right)_0^e \sigma_{xz}^{(1)} d\ell \right\}.$$

$$A_{20'2} = \left\{ M_R (L - \ell_{CG}) \left[\varphi_{xz}^{(2)}(eG) - \ell_R \, \sigma_{xz}^{(2)}(eG) \right] \right\} s^2$$

$$+ \left\{ - \frac{\rho U_0^2 S_5 b_2}{2} \left[\int_0^L \left(\frac{\partial C_m(\ell)}{\partial w} \right)_0^e \varphi_{xz}^{(2)} d\ell - \int_0^L \left(\frac{\partial C_m(\ell)}{\partial \dot{w}} \right)_0^e U_0 \, \sigma_{xz}^{(2)} d\ell \right] \right\} s$$

$$+ \left\{ -\ell_e (T_s + T_c) \sigma_{xz}^{(2)}(eG) - (T_s + T_c) \varphi_{xz}^{(2)}(eG) + \frac{\rho U_0^3 S_5 b_2}{2} \int_0^L \left(\frac{\partial C_m(\ell)}{\partial w} \right)_0^e \sigma_{xz}^{(2)} d\ell \right\}.$$

$$A_{20'3} = \left\{ M_R (L - \ell_{CG}) \left[\varphi_{xz}^{(3)}(eG) - \ell_R \, \sigma_{xz}^{(3)}(eG) \right] \right\} s^2$$

$$+ \left\{ - \frac{\rho U_0^2 S_5 b_2}{2} \left[\int_0^L \left(\frac{\partial C_m(\ell)}{\partial w} \right)_0^e \varphi_{xz}^{(3)} d\ell - \int_0^L \left(\frac{\partial C_m(\ell)}{\partial \dot{w}} \right)_0^e U_0 \, \sigma_{xz}^{(3)} d\ell \right] \right\} s$$

$$+ \left\{ -\ell_e (T_s + T_c) \sigma_{xz}^{(3)}(eG) - (T_s + T_c) \varphi_{xz}^{(3)}(eG) + \frac{\rho U_0^3 S_5 b_2}{2} \int_0^L \left(\frac{\partial C_m(\ell)}{\partial w} \right)_0^e \sigma_{xz}^{(3)} d\ell \right\}.$$

$$A_{20'4} = \left\{ M_R (L - \ell_{CG}) \left[\varphi_{xz}^{(4)}(eG) - \ell_R \, \sigma_{xz}^{(4)}(eG) \right] \right\} s^2$$

$$+ \left\{ - \frac{\rho U_0^2 S_5 b_2}{2} \left[\int_0^L \left(\frac{\partial C_m(\ell)}{\partial w} \right)_0^e \varphi_{xz}^{(4)} d\ell - \int_0^L \left(\frac{\partial C_m(\ell)}{\partial \dot{w}} \right)_0^e U_0 \, \sigma_{xz}^{(4)} d\ell \right] \right\} s$$

$$+ \left\{ -\ell_e (T_s + T_c) \sigma_{xz}^{(4)}(eG) - (T_s + T_c) \varphi_{xz}^{(4)}(eG) + \frac{\rho U_0^3 S_5 b_2}{2} \int_0^L \left(\frac{\partial C_m(\ell)}{\partial w} \right)_0^e \sigma_{xz}^{(4)} d\ell \right\}.$$

$$A_{20'5} = \left\{ M_R (L - \ell_{CG}) \left[\varphi_{xz}^{(5)}(eG) - \ell_R \, \sigma_{xz}^{(5)}(eG) \right] \right\} s^2$$

$$+ \left\{ - \frac{\rho U_0^2 S_5 b_2}{2} \left[\int_0^L \left(\frac{\partial C_m(\ell)}{\partial w} \right)_0^e \varphi_{xz}^{(5)} d\ell - \int_0^L \left(\frac{\partial C_m(\ell)}{\partial \dot{w}} \right)_0^e U_0 \, \sigma_{xz}^{(5)} d\ell \right] \right\} s$$

$$+ \left\{ -\ell_e (T_s + T_c) \sigma_{xz}^{(5)}(eG) - (T_s + T_c) \varphi_{xz}^{(5)}(eG) + \frac{\rho U_0^3 S_5 b_2}{2} \int_0^L \left(\frac{\partial C_m(\ell)}{\partial w} \right)_0^e \sigma_{xz}^{(5)} d\ell \right\}.$$

$$A_{20'6} = \left\{\frac{-\rho U_0^2 S_5 b_2}{2} \int_0^L \left(\frac{\partial C_m(\ell)}{\partial v}\right)_0^e \varphi_{xy}^{(1)} d\ell \right\} s + \left\{\frac{\rho U_0^3 S_5 b_2}{2} \int_0^L \left(\frac{\partial C_m(\ell)}{\partial v}\right)_0^e \sigma_{xy}^{(1)} d\ell \right\}$$

$$A_{20'7} = \left\{\frac{-\rho U_0^2 S_5 b_2}{2} \int_0^L \left(\frac{\partial C_m(\ell)}{\partial v}\right)_0^e \varphi_{xy}^{(2)} d\ell \right\} s + \left\{\frac{\rho U_0^3 S_5 b_2}{2} \int_0^L \left(\frac{\partial C_m(\ell)}{\partial v}\right)_0^e \sigma_{xy}^{(2)} d\ell \right\}.$$

$$A_{20'8} = \left\{\frac{-\rho U_0^2 S_5 b_2}{2} \int_0^L \left(\frac{\partial C_m(\ell)}{\partial v}\right)_0^e \varphi_{xy}^{(3)} d\ell \right\} s + \left\{\frac{\rho U_0^3 S_5 b_2}{2} \int_0^L \left(\frac{\partial C_m(\ell)}{\partial v}\right)_0^e \sigma_{xy}^{(3)} d\ell \right\}$$

$$A_{20'9} = \left\{\frac{-\rho U_0^2 S_5 b_2}{2} \int_0^L \left(\frac{\partial C_m(\ell)}{\partial v}\right)_0^e \varphi_{xy}^{(4)} d\ell \right\} s + \left\{\frac{\rho U_0^3 S_5 b_2}{2} \int_0^L \left(\frac{\partial C_m(\ell)}{\partial v}\right)_0^e \sigma_{xy}^{(4)} d\ell \right\}.$$

$$A_{20'10} = \left\{\frac{-\rho U_0^2 S_5 b_2}{2} \int_0^L \left(\frac{\partial C_m(\ell)}{\partial v}\right)_0^e \varphi_{xy}^{(5)} d\ell \right\} s + \left\{\frac{\rho U_0^3 S_5 b_2}{2} \int_0^L \left(\frac{\partial C_m(\ell)}{\partial v}\right)_0^e \sigma_{xy}^{(5)} d\ell \right\}.$$

$$A_{20'11} = \left\{\frac{-\rho U_0^2 S_5 b_2}{2} \int_0^L \left(\frac{\partial C_m(\ell)}{\partial p}\right)_0^e \varphi_{zy}^{(1)} d\ell \right\} s.$$

$$A_{20'12} = \left\{\frac{-\rho U_0^2 S_5 b_2}{2} \int_0^L \left(\frac{\partial C_m(\ell)}{\partial p}\right)_0^e \varphi_{zy}^{(2)} d\ell \right\} s.$$

$$A_{20'13} = \left\{\frac{-\rho U_0^2 S_5 b_2}{2} \int_0^L \left(\frac{\partial C_m(\ell)}{\partial p}\right)_0^e \varphi_{zy}^{(3)} d\ell \right\} s.$$

$$A_{20'14} = \left\{\frac{-\rho U_0^2 S_5 b_2}{2} \int_0^L \left(\frac{\partial C_m(\ell)}{\partial p}\right)_0^e \varphi_{zy}^{(4)} d\ell \right\} s.$$

$$A_{20'15} = \left\{\frac{-\rho U_0^2 S_5 b_2}{2} \int_0^L \left(\frac{\partial C_m(\ell)}{\partial p}\right)_0^e \varphi_{zy}^{(5)} d\ell \right\} s.$$

$$A_{20'16} = \left\{\frac{-\rho U_0^2 S_5 b_2}{2} \left[\left(\frac{\partial C_m}{\partial u}\right)_0^e + \frac{2}{U_0} (C_m)_0^e \right]\right\}.$$

$$A_{20'17} = \left\{\frac{-\rho U_0^2 S_5 b_2}{2} \left(\frac{\partial C_m}{\partial v}\right)_0^e \right\}.$$

$$A_{20'18} = \left\{M_R (L - \ell_{CG}) - \frac{\rho U_0^2 S_5 b_2}{2} \left(\frac{\partial C_m}{\partial \dot{w}}\right)_0^e \right\} s + \left\{\frac{-\rho U_0^2 S_5 b_2}{2} \left(\frac{\partial C_m}{\partial w}\right)_0^e \right\}.$$

$$A_{20'19} = \left\{2 I_{xz} P_0 + (I_{xx} - I_{zz}) R_0 - \frac{\rho U_0^2 S_5 b_2}{2} \left(\frac{\partial C_m}{\partial p}\right)_0^e \right\} s$$

$$A_{20'20} = \left\{ M_R (L - \ell_{CG})^2 + I_{yy} \right\} s^2 + \left\{ \frac{-\rho U_0^2 S_5 b_2}{2} \left(\frac{\partial C_m}{\partial q}\right)_0^e \right\} s.$$

$$A_{20'21} = \left\{ (I_{xx} - I_{zz}) P_0 - 2 I_{xz} R_0 - \frac{\rho U_0^2 S_5 b_2}{2} \left(\frac{\partial C_m}{\partial r}\right)_0^e \right\} s.$$

$$A_{20'22} = \left\{ -M_{c_1} \right\} s^2 + \left\{ -K_1 X_1 - \frac{M_{l_1}(T_s + T_c - D)}{M_t} \right\}.$$

$$A_{20'23} = \left\{ -M_{c_2} \right\} s^2 + \left\{ -K_2 X_2 - \frac{M_{l_2}(T_s + T_c - D)}{M_t} \right\}.$$

$$A_{20'24} = \left\{ -M_{c_3} \right\} s^2 + \left\{ -K_3 X_3 - \frac{M_{l_3}(T_s + T_c - D)}{M_t} \right\}.$$

$$A_{20'25} = \left\{ -M_{c_4} \right\} s^2 + \left\{ -K_4 X_4 - \frac{M_{l_4}(T_s + T_c - D)}{M_t} \right\}.$$

$$A_{20'26} \cdots A_{20'29} = 0.$$

$$A_{20'30} = \left\{ M_R \ell_R \right\} s^2 + \left\{ \ell_e \; T_c \right\}.$$

$$A_{20'31} \cdots A_{20'40} = 0.$$

$$A_{21'1} = \left\{ \frac{-\rho U_0^2 S_6 b_3}{2} \int_0^L \left(\frac{\partial C_n(\ell)}{\partial w}\right)_0^e \varphi_{xz}^{(1)} d\ell \right\} s + \left\{ \frac{\rho U_0^3 S_6 b_3}{2} \int_0^L \left(\frac{\partial C_n(\ell)}{\partial w}\right)_0^e \sigma_{xz}^{(1)} d\ell \right\}.$$

$$A_{21'2} = \left\{ \frac{-\rho U_0^2 S_6 b_3}{2} \int_0^L \left(\frac{\partial C_n(\ell)}{\partial w}\right)_0^e \varphi_{xz}^{(2)} d\ell \right\} s + \left\{ \frac{\rho U_0^3 S_6 b_3}{2} \int_0^L \left(\frac{\partial C_n(\ell)}{\partial w}\right)_0^e \sigma_{xz}^{(2)} d\ell \right\}.$$

$$A_{21'3} = \left\{ \frac{-\rho U_0^2 S_6 b_3}{2} \int_0^L \left(\frac{\partial C_n(\ell)}{\partial w}\right)_0^e \varphi_{xz}^{(3)} d\ell \right\} s + \left\{ \frac{\rho U_0^3 S_6 b_3}{2} \int_0^L \left(\frac{\partial C_n(\ell)}{\partial w}\right)_0^e \sigma_{xz}^{(3)} d\ell \right\}.$$

$$A_{21'4} = \left\{ \frac{-\rho U_0^2 S_6 b_3}{2} \int_0^L \left(\frac{\partial C_n(\ell)}{\partial w}\right)_0^e \varphi_{xz}^{(4)} d\ell \right\} s + \left\{ \frac{\rho U_0^3 S_6 b_3}{2} \int_0^L \left(\frac{\partial C_n(\ell)}{\partial w}\right)_0^e \sigma_{xz}^{(4)} d\ell \right\}.$$

$$A_{21'5} = \left\{ \frac{-\rho U_0^2 S_6 b_3}{2} \int_0^L \left(\frac{\partial C_n(\ell)}{\partial w}\right)_0^e \varphi_{xz}^{(5)} d\ell \right\} s + \left\{ \frac{\rho U_0^3 S_6 b_3}{2} \int_0^L \left(\frac{\partial C_n(\ell)}{\partial w}\right)_0^e \sigma_{xz}^{(5)} d\ell \right\}.$$

$$A_{21'6} = \left\{ M_R \,(L - \ell_{CG}) \left[\varphi_{xy}^{(1)}(eG) - \ell_R \, \sigma_{xy}^{(1)}(eG) \right] \right\} s^2$$

$$+ \left\{ \frac{-\rho U_0^2 S_6 b_3}{2} \left[\int_0^L \left(\frac{\partial C_n(\ell)}{\partial v} \right)_0^e \varphi_{xy}^{(1)} d\ell - \int_0^L \left(\frac{\partial C_n(\ell)}{\partial \dot{v}} \right)_0^e U_0 \, \sigma_{xy}^{(1)} d\ell \right] \right\} s$$

$$+ \left\{ \ell_e (T_s + T_c) \, \sigma_{xy}^{(1)}(eG) + (T_s + T_c) \, \varphi_{xy}^{(1)}(eG) + \frac{\rho U_0^3 S_6 b_3}{2} \int_0^L \left(\frac{\partial C_n(\ell)}{\partial v} \right)_0^e \sigma_{xy}^{(1)} d\ell \right\}.$$

$$A_{21'7} = \left\{ M_R \,(L - \ell_{CG}) \left[\varphi_{xy}^{(2)}(eG) - \ell_R \, \sigma_{xy}^{(2)}(eG) \right] \right\} s^2$$

$$+ \left\{ - \frac{\rho U_0^2 S_6 b_3}{2} \left[\int_0^L \left(\frac{\partial C_n(\ell)}{\partial v} \right)_0^e \varphi_{xy}^{(2)} d\ell - \int_0^L \left(\frac{\partial C_n(\ell)}{\partial \dot{v}} \right)_0^e U_0 \, \sigma_{xy}^{(2)} d\ell \right] \right\} s$$

$$+ \left\{ \ell_e (T_s + T_c) \, \sigma_{xy}^{(2)}(eG) + (T_s + T_c) \, \varphi_{xy}^{(2)}(eG) + \frac{\rho U_0^3 S_6 b_3}{2} \int_0^L \left(\frac{\partial C_n(\ell)}{\partial v} \right)_0^e \sigma_{xy}^{(2)} d\ell \right\}.$$

$$A_{21'8} = \left\{ M_R \,(L - \ell_{CG}) \left[\varphi_{xy}^{(3)}(eG) - \ell_R \, \sigma_{xy}^{(3)}(eG) \right] \right\} s^2$$

$$+ \left\{ - \frac{\rho U_0^2 S_6 b_3}{2} \left[\int_0^L \left(\frac{\partial C_n(\ell)}{\partial v} \right)_0^e \varphi_{xy}^{(3)} d\ell - \int_0^L \left(\frac{\partial C_n(\ell)}{\partial \dot{v}} \right)_0^e U_0 \, \sigma_{xy}^{(3)} d\ell \right] \right\} s$$

$$+ \left\{ \ell_e (T_s + T_c) \, \sigma_{xy}^{(3)}(eG) + (T_s + T_c) \, \varphi_{xy}^{(3)}(eG) + \frac{\rho U_0^3 S_6 b_3}{2} \int_0^L \left(\frac{\partial C_n(\ell)}{\partial v} \right)_0^e \sigma_{xy}^{(3)} d\ell \right\}.$$

$$A_{21'9} = \left\{ M_R \,(L - \ell_{CG}) \left[\varphi_{xy}^{(4)}(eG) - \ell_R \, \sigma_{xy}^{(4)}(eG) \right] \right\} s^2$$

$$+ \left\{ - \frac{\rho U_0^2 S_6 b_3}{2} \left[\int_0^L \left(\frac{\partial C_n(\ell)}{\partial v} \right)_0^e \varphi_{xy}^{(4)} d\ell - \int_0^L \left(\frac{\partial C_n(\ell)}{\partial \dot{v}} \right)_0^e U_0 \, \sigma_{xy}^{(4)} d\ell \right] \right\} s$$

$$+ \left\{ \ell_e (T_s + T_c) \, \sigma_{xy}^{(4)}(eG) + (T_s + T_c) \, \varphi_{xy}^{(4)}(eG) + \frac{\rho U_0^3 S_6 b_3}{2} \int_0^L \left(\frac{\partial C_n(\ell)}{\partial v} \right)_0^e \sigma_{xy}^{(4)} d\ell \right\}.$$

$$A_{21'10} = \left\{ M_R \,(L - \ell_{CG}) \left[\varphi_{xy}^{(5)}(eG) - \ell_R \, \sigma_{xy}^{(5)}(eG) \right] \right\} s^2$$

$$+ \left\{ - \frac{\rho U_0^2 S_6 b_3}{2} \left[\int_0^L \left(\frac{\partial C_n(\ell)}{\partial v} \right)_0^e \varphi_{xy}^{(5)} d\ell - \int_0^L \left(\frac{\partial C_n(\ell)}{\partial \dot{v}} \right)_0^e U_0 \, \sigma_{xy}^{(5)} d\ell \right] \right\} s$$

$$+ \left\{ \ell_e (T_s + T_c) \, \sigma_{xy}^{(5)}(eG) + (T_s + T_c) \, \varphi_{xy}^{(5)}(eG) + \frac{\rho U_0^3 S_6 b_3}{2} \int_0^L \left(\frac{\partial C_n(\ell)}{\partial v} \right)_0^e \sigma_{xy}^{(5)} d\ell \right\}$$

$$A_{21'11} = \left\{ \frac{-\rho U_0^2 S_6 b_3}{2} \int_0^L \left(\frac{\partial C_n(\ell)}{\partial p} \right)_0^e \varphi_{zy}^{(1)} \, d\ell \right\} s.$$

$$A_{21'12} = \left\{ \frac{-\rho U_0^2 S_6 b_3}{2} \int_0^L \left(\frac{\partial C_n(\ell)}{\partial p} \right)_0^e \varphi_{zy}^{(2)} \, d\ell \right\} s.$$

$$A_{21'13} = \left\{ \frac{-\rho U_0^2 S_6 b_3}{2} \int_0^L \left(\frac{\partial C_n(\ell)}{\partial p} \right)_0^e \varphi_{zy}^{(3)} \, d\ell \right\} s.$$

$$A_{21'14} = \left\{ \frac{-\rho U_0^2 S_6 b_3}{2} \int_0^L \left(\frac{\partial C_n(\ell)}{\partial p} \right)_0^e \varphi_{zy}^{(4)} \, d\ell \right\} s$$

$$A_{21'15} = \left\{ \frac{-\rho U_0^2 S_6 b_3}{2} \int_0^L \left(\frac{\partial C_n(\ell)}{\partial p} \right)_0^e \varphi_{zy}^{(5)} \, d\ell \right\} s.$$

$$A_{21'16} = \left\{ \frac{-\rho U_0^2 S_6 b_3}{2} \left[\left(\frac{\partial C_n}{\partial u} \right)_0^e + \frac{2}{U_0} (C_n)_0^e \right] \right\}.$$

$$A_{21'17} = \left\{ M_R (L - \ell_{CG}) - \frac{\rho U_0^2 S_6 b_3}{2} \left(\frac{\partial C_n}{\partial \dot{v}} \right)_0^e \right\} s + \left\{ \frac{-\rho U_0^2 S_6 b_3}{2} \left(\frac{\partial C_n}{\partial v} \right)_0^e \right\}.$$

$$A_{21'18} = \left\{ \frac{-\rho U_0^2 S_6 b_3}{2} \left(\frac{\partial C_n}{\partial w} \right)_0^e \right\}.$$

$$A_{21'19} = \left\{ -I_{xz} \right\} s^2 + \left\{ (I_{yy} - I_{xx}) Q_0 - \frac{\rho U_0^2 S_6 b_3}{2} \left(\frac{\partial C_n}{\partial p} \right)_0^e \right\} s.$$

$$A_{21'20} = \left\{ R_0 I_{xz} + (I_{yy} - I_{xx}) P_0 - \frac{\rho U_0^2 S_6 b_3}{2} \left(\frac{\partial C_n}{\partial q} \right)_0^e \right\} s.$$

$$A_{21'21} = \left\{ I_{zz} - M_R (L - \ell_{CG})^2 \right\} s^2 + \left\{ I_{xz} Q_0 - \frac{\rho U_0^2 S_6 b_3}{2} \left(\frac{\partial C_n}{\partial r} \right)_0^e \right\} s.$$

$$A_{21'22} \cdots A_{21'25} = 0.$$

$$A_{21'26} = \left\{ M_{C_1} \right\} s^2 + \left\{ K_1 X_1 + \frac{M_{l_1}(T_s + T_c - D)}{M_t} \right\}.$$

$$A_{21'27} = \left\{ M_{C_2} \right\} s^2 + \left\{ K_2 X_2 + \frac{M_{l_2}(T_s + T_c - D)}{M_t} \right\}.$$

$$A_{21'28} = \left\{ M_{C_3} \right\} s^2 + \left\{ K_3 X_3 + \frac{M_{l_3}(T_s + T_c - D)}{M_t} \right\}.$$

$$A_{21'29} = \left\{ M_{C_4} \right\} s^2 + \left\{ K_4 X_4 + \frac{M_{l_4}(T_s + T_c - D)}{M_t} \right\}.$$

$A_{21,30} = 0.$

$A_{21,31} = \{ M_R \ell_R (L - \ell_{CG}) \} s^2 + \{ -T_c \ell_e \}.$

$A_{21,32} \cdots\cdots A_{21,40} = 0.$

$A_{22,1} = \{ -M_{I_1} \varphi^{(1)}_{xz} (X_1) \} s^2.$

$A_{22,2} = \{ -M_{I_1} \varphi^{(2)}_{xz} (X_1) \} s^2.$

$A_{22,3} = \{ -M_{I_1} \varphi^{(3)}_{xz} (X_1) \} s^2$

$A_{22,4} = \{ -M_{I_1} \varphi^{(4)}_{xz} (X_1) \} s^2$

$A_{22,5} = \{ -M_{I_1} \varphi^{(5)}_{xz} (X_1) \} s^2$

$A_{22,6} \cdots\cdots A_{22,17} = 0.$

$A_{22,18} = \{ -M_{I_1} \} s.$

$A_{22,19} = 0.$

$A_{22,20} = \{ M_{I_1} X_1 \} s^2$

$A_{22,21} = 0.$

$A_{22,22} = \{ M_{I_1} \} s^2 + \{ 2\xi_1 \omega_1 M_{I_1} \} s + \{ K_1 \}.$

$A_{22,23} \cdots\cdots A_{22,40} = 0.$

$A_{23,1} = \{ -M_{I_2} \varphi^{(1)}_{xz} (X_2) \} s^2.$

$A_{23,2} = \{ -M_{I_2} \varphi^{(2)}_{xz} (X_2) \} s^2.$

$A_{23,3} = \{ -M_{I_2} \varphi^{(3)}_{xz} (X_2) \} s^2$

$A_{23,4} = \{ -M_{I_2} \varphi^{(4)}_{xz} (X_2) \} s^2$

$A_{23,5} = \{ -M_{I_2} \varphi^{(5)}_{xz} (X_2) \} s^2$

$A_{23,6} \cdots\cdots A_{23,17} = 0.$

$$A_{23'18} = \left\{ - M_{1_2} \right\} s .$$

$$A_{23'19} = 0 .$$

$$A_{23'20} = \left\{ M_{1_2} X_2 \right\} s^2$$

$$A_{23'21} \cdots A_{23'22} = 0$$

$$A_{23'23} = \left\{ M_{1_2} \right\} s^2 + \left\{ 2\xi_2 \, \omega_2 M_{1_2} \right\} s + \left\{ K_2 \right\} .$$

$$A_{23'24} \cdots\cdots A_{23'40} = 0 .$$

$$A_{24'1} = \left\{ - M_{1_3} \; \varphi_{xz}^{(1)} (X_3) \right\} s^2 .$$

$$A_{24'2} = \left\{ - M_{1_3} \; \varphi_{xz}^{(2)} (X_3) \right\} s^2 .$$

$$A_{24'3} = \left\{ - M_{1_3} \; \varphi_{xz}^{(3)} (X_3) \right\} s^2 .$$

$$A_{24'4} = \left\{ - M_{1_3} \; \varphi_{xz}^{(4)} (X_3) \right\} s^2$$

$$A_{24'5} = \left\{ - M_{1_3} \; \varphi_{xz}^{(5)} (X_3) \right\} s^2$$

$$A_{24'6} \cdots\cdots A_{24'17} = 0 .$$

$$A_{24'18} = \left\{ - M_{1_3} \right\} s .$$

$$A_{24'19} = 0 .$$

$$A_{24\,20} = \left\{ M_{1_3} X_3 \right\} s^2$$

$$A_{24'21} \cdots A_{24'23} = 0 .$$

$$A_{24'24} = \left\{ M_{1_3} \right\} s^2 + \left\{ 2\xi_3 \; \omega_3 M_{1_3} \right\} s + \left\{ K_3 \right\} .$$

$$A_{24'25} \cdots A_{24'40} = 0 .$$

$$A_{25'1} = \left\{ - M_{1_4} \; \varphi_{xz}^{(1)} (X_4) \right\} s^2 .$$

$$A_{25'2} = \left\{ - M_{1_4} \; \varphi_{xz}^{(2)} (X_4) \right\} s^2$$

$$A_{25'3} = \left\{ - M_{1_4} \; \varphi_{xz}^{(3)} (X_4) \right\} s^2$$

$$A_{25'4} = \left\{ - M_{1_4} \; \varphi_{xz}^{(4)} \; (X_4) \right\} s^2 .$$

$$A_{25'5} = \left\{ - M_{1_4} \; \varphi_{xz}^{(5)} \; (X_4) \right\} s^2 .$$

$$A_{25'6} \cdots\cdots A_{25'17} = 0 .$$

$$A_{25'18} = \left\{ - M_{1_4} \right\} s .$$

$$A_{25'19} = 0 .$$

$$A_{25'20} = \left\{ M_{1_4} X_4 \right\} s^2$$

$$A_{25'21} \cdots\cdots A_{25'24} = 0 .$$

$$A_{25'25} = \left\{ M_{1_4} \right\} s^2 + \left\{ 2 \xi_4 \; \omega_4 \; M_{1_4} \right\} s + \left\{ K_4 \right\} .$$

$$A_{25'26} \cdots\cdots A_{25'40} = 0 .$$

$$A_{26'1} \cdots\cdots A_{26'5} = 0 .$$

$$A_{26'6} = \left\{ - M_{1_1} \; \varphi_{xy}^{(1)} (X_1) \right\} s^2 .$$

$$A_{26'7} = \left\{ - M_{1_1} \; \varphi_{xy}^{(2)} (X_1) \right\} s^2$$

$$A_{26'8} = \left\{ - M_{1_1} \; \varphi_{xy}^{(3)} \; (X_1) \right\} s^2$$

$$A_{26'9} = \left\{ - M_{1_1} \; \varphi_{xy}^{(4)} \; (X_1) \right\} s^2 .$$

$$A_{26'10} = \left\{ - M_{1_1} \; \varphi_{xy}^{(5)} \; (X_1) \right\} s^2 .$$

$$A_{26'11} \cdots\cdots A_{26'16} = 0 .$$

$$A_{26'17} = \left\{ - M_{1_1} \right\} s .$$

$$A_{26'18} \cdots\cdots A_{26'20} = 0 .$$

$$A_{26'21} = \left\{ - M_{1_1} X_1 \right\} s^2$$

$$A_{26'22} \cdots\cdots A_{26'25} = 0 .$$

$$A_{26'26} = \left\{ M_{1_1} \right\} s^2 + \left\{ 2 \xi_1 \; \omega_1 \; M_{1_1} \right\} s + \left\{ K_1 \right\} .$$

$$A_{26'27} \cdots\cdots A_{26'40} = 0 .$$

$A_{27'1} \cdots\cdots A_{27'5} = 0.$

$A_{27'6} = \left\{ -M_{1_2} \varphi^{(1)}_{xy}(X_2) \right\} s^2.$

$A_{27'7} = \left\{ -M_{1_2} \varphi^{(2)}_{xy}(X_2) \right\} s^2$

$A_{27'8} = \left\{ -M_{1_2} \varphi^{(3)}_{xy}(X_2) \right\} s^2$

$A_{27'9} = \left\{ -M_{1_2} \varphi^{(4)}_{xy}(X_2) \right\} s^2$

$A_{27'10} = \left\{ -M_{1_2} \varphi^{(5)}_{xy}(X_2) \right\} s^2.$

$A_{27'11} \cdots\cdots A_{27'16} = 0.$

$A_{27'17} = \left\{ -M_{1_2} \right\} s.$

$A_{27'18} \cdots\cdots A_{27'20} = 0.$

$A_{27'21} = \left\{ -M_{1_2} X_2 \right\} s^2$

$A_{27'22} \cdots\cdots A_{27'26} = 0.$

$A_{27'27} = \left\{ M_{1_2} \right\} s^2 + \left\{ 2\xi_2 \omega_2 M_{1_2} \right\} s + \left\{ K_2 \right\}.$

$A_{27'27} \cdots\cdots A_{27'40} = 0.$

$A_{28'1} \cdots\cdots A_{28'5} = 0.$

$A_{28'6} = \left\{ -M_{1_3} \varphi^{(1)}_{xy}(X_3) \right\} s^2.$

$A_{28'7} = \left\{ -M_{1_3} \varphi^{(2)}_{xy}(X_3) \right\} s^2$

$A_{28'8} = \left\{ -M_{1_3} \varphi^{(3)}_{xy}(X_3) \right\} s^2$

$A_{28'9} = \left\{ -M_{1_3} \varphi^{(4)}_{xy}(X_3) \right\} s^2$

$A_{28'10} = \left\{ -M_{1_3} \varphi^{(5)}_{xy}(X_3) \right\} s^2.$

$A_{28'11} \cdots\cdots A_{28'16} = 0.$

$A_{28'17} = \left\{ -M_{1_3} \right\} s.$

$A_{28'18} \cdots\cdots A_{28'20} = 0.$

$$A_{28'21} = \{-M_{I_3} X_3\} s^2$$

$$A_{28'22} \cdots A_{28'27} = 0.$$

$$A_{28'28} = \{M_{I_3}\} s^2 + \{2 \xi_3 \omega_3 M_{I_3}\} s + \{K_3\}.$$

$$A_{28'29} \cdots A_{28'40} = 0.$$

$$A_{29'1} \cdots A_{29'5} = 0.$$

$$A_{29'6} = \{-M_{I_4} \varphi_{xy}^{(1)}(X_4)\} s^2$$

$$A_{29'7} = \{-M_{I_4} \varphi_{xy}^{(2)}(X_4)\} s^2$$

$$A_{29'8} = \{-M_{I_4} \varphi_{xy}^{(3)}(X_4)\} s^2$$

$$A_{29'9} = \{-M_{I_4} \varphi_{xy}^{(4)}(X_4)\} s^2$$

$$A_{29'10} = \{-M_{I_4} \varphi_{xy}^{(5)}(X_4)\} s^2$$

$$A_{29'11} \cdots A_{29'16} = 0.$$

$$A_{29'17} = \{-M_{I_4}\} s.$$

$$A_{29'18} \cdots A_{29'20} = 0.$$

$$A_{29'21} = \{-M_{I_4} X_4\} s^2$$

$$A_{29'22} \cdots A_{29'28} = 0.$$

$$A_{29'29} = \{M_{I_4}\} s^2 + \{2 \xi_4 \omega_4 M_{I_4}\} s + \{K_4\}.$$

$$A_{29'30} \cdots A_{29'40} = 0.$$

$$A_{30'1} = \{M_R \ell_R [\varphi_{xz}^{(1)}(eG) - \ell_R \sigma_{xz}^{(1)}(eG)]\} s^2 + \{-M_R \ell_R \dot{U}_0 \sigma_{xz}^{(1)}(eG)\}.$$

$$A_{30'2} = \{M_R \ell_R [\varphi_{xz}^{(2)}(eG) - \ell_R \sigma_{xz}^{(2)}(eG)]\} s^2 + \{-M_R \ell_R \dot{U}_0 \sigma_{xz}^{(2)}(eG)\}.$$

$$A_{30'3} = \{M_R \ell_R [\varphi_{xz}^{(3)}(eG) - \ell_R \sigma_{xz}^{(3)}(eG)]\} s^2 + \{-M_R \ell_R \dot{U}_0 \sigma_{xz}^{(3)}(eG)\}.$$

$$A_{30'4} = \{M_R \ell_R [\varphi_{xz}^{(4)}(eG) - \ell_R \sigma_{xz}^{(4)}(eG)]\} s^2 + \{-M_R \ell_R \dot{U}_0 \sigma_{xz}^{(4)}(eG)\}.$$

$$A_{30'5} = \{M_R \ell_R [\varphi_{xz}^{(5)}(eG) - \ell_R \sigma_{xz}^{(5)}(eG)]\} s^2 + \{-M_R \ell_R U_0 \sigma_{xz}^{(5)}(eG)\}.$$

$A_{30,6} \cdots\cdots A_{30,17} = 0.$

$A_{30,18} = \{ M_R \ell_R \} s$

$A_{30,19} = 0.$

$A_{30,20} = \{ M_R \ell_R (L - \ell_{CG}) \} s^2$

$A_{30,21} \cdots\cdots A_{30,29} = 0.$

$A_{30,30} = \{ -\overline{K}_2 \} s^3 + \{ M_R \ell_R^2 \dot{U}_0 - \overline{K}_3 \} s^2 + \{ \overline{C}_{f_{xz}} - \overline{K}_4 \} s + \{ M_R \ell_R \dot{U}_0 - \overline{K}_5 \}.$

$A_{30,31} = 0.$

$A_{30,32} = \{ -\overline{K}_1 \}.$

$A_{30,33} \cdots\cdots A_{30,40} = 0.$

$A_{31,1} \cdots\cdots A_{31,5} = 0.$

$A_{31,6} = \{ M_R \ell_R [\varphi_{xy}^{(1)}(eG) - \ell_R \; \sigma_{xy}^{(1)}(eG)] \} s^2.$

$A_{31,7} = \{ M_R \ell_R [\varphi_{xy}^{(2)}(eG) - \ell_R \; \sigma_{xy}^{(2)}(eG)] \} s^2.$

$A_{31,8} = \{ M_R \ell_R [\varphi_{xy}^{(3)}(eG) - \ell_R \; \sigma_{xy}^{(3)}(eG)] \} s^2$

$A_{31,9} = \{ M_R \ell_R [\varphi_{xy}^{(4)}(eG) - \ell_R \; \sigma_{xy}^{(4)}(eG)] \} s^2.$

$A_{31,10} = \{ M_R \ell_R [\varphi_{xy}^{(5)}(eG) - \ell_R \; \sigma_{xy}^{(5)}(eG)] \} s^2$

$A_{31,11} \cdots\cdots A_{31,15} = 0.$

$A_{31,16} = \{ M_R \ell_R \} s.$

$A_{31,17} = \{ M_R \ell_R \} s.$

$A_{31,18} \cdots\cdots A_{31,20} = 0.$

$A_{31,21} = \{ -M_R \ell_R (L - \ell_{CG}) \} s^2$

$A_{31,22} \cdots\cdots A_{31,30} = 0.$

$A_{31,31} = \{ -\overline{K}_7 \} s^3 + \{ M_R \ell_R^2 - \overline{K}_8 \} s^2 + \{ \overline{C}_{f_{xy}} - \overline{K}_9 \} s + \{ -\overline{K}_{10} \}$

$A_{31,32} = 0.$

$A_{31,33} = \{ -\overline{K}_6 \}.$

$A_{31,34} \cdots\cdots A_{31,40} = 0.$

$A_{32,1} \quad \cdots\cdots \quad A_{32,31} = 0.$

$$A_{32,32} = \left\{ \left(\frac{\tau_f}{K_A\, K_I}\right)_{xz} \right\} s^2 + \left\{ \left(\frac{1}{K_A\, K_I}\right)_{xz} \right\} s .$$

$A_{32,33} \quad \cdots\cdots \quad A_{32,34} = 0.$

$$A_{32,35} = \left\{ \left(\frac{1}{K_I}\right)_{xz} + \tau_{f_{xz}} \right\} s + \left\{ 1 \right\}.$$

$A_{32,36} \quad \cdots\cdots \quad A_{32,37} = 0.$

$$A_{32,38} = \left\{ \left(\frac{1}{K_I}\right)_{xz} + \tau_{f_{xz}} \right\} s + \left\{ 1 \right\}.$$

$A_{32,39} \quad \cdots\cdots \quad A_{32,40} = 0.$

$A_{33,1} \quad \cdots\cdots \quad A_{33,32} = 0.$

$$A_{33,33} = \left\{ \left(\frac{\tau_f}{K_A\, K_I}\right)_{xy} \right\} s^2 + \left\{ \left(\frac{1}{K_A\, K_I}\right)_{xy} \right\} s .$$

$A_{33,34} \quad \cdots\cdots \quad A_{33,35} = 0.$

$$A_{33,36} = \left\{ \left(\frac{1}{K_I}\right)_{xy} + \tau_{f_{xy}} \right\} s + \left\{ 1 \right\}.$$

$A_{33,37} \quad \cdots\cdots \quad A_{33,38} = 0.$

$$A_{33,39} = \left\{ \left(\frac{1}{K_I}\right)_{xy} + \tau_{f_{xy}} \right\} s + \left\{ 1 \right\}$$

$A_{33,40} = 0.$

$A_{34,1} \quad \cdots\cdots \quad A_{34,33} = 0.$

$$A_{34,34} = \left\{ \left(\frac{\tau_f}{K_A\, K_I}\right)_{zy} \right\} s^2 + \left\{ \left(\frac{1}{K_A\, K_I}\right)_{zy} \right\} s .$$

$A_{34,35} \quad \cdots\cdots \quad A_{34,36} = 0.$

$$A_{34,37} = \left\{ \left(\frac{1}{K_I}\right)_{zy} + \tau_{f_{zy}} \right\} s + \left\{ 1 \right\}$$

$A_{34,38} \quad \cdots\cdots \quad A_{34,39} = 0.$

$$A_{34,40} = \left\{ \left(\frac{1}{K_I}\right)_{zy} + \tau_{f_{zy}} \right\} s + \left\{ 1 \right\}$$

$$A_{35,1} = \left\{ -\left(\omega_{RG}\right)^2_{xz} \quad K_{R_{xz}} \quad \sigma^{(1)}_{xz}(RG) \right\} s.$$

$$A_{35,2} = \left\{ -\left(\omega_{RG}\right)^2_{xz} \quad K_{R_{xz}} \quad \sigma^{(2)}_{xz}(RG) \right\} s.$$

$$A_{35,3} = \left\{ -\left(\omega_{RG}\right)^2_{xz} \quad K_{R_{xz}} \quad \sigma^{(3)}_{xz}(RG) \right\} s.$$

$$A_{35,4} = \left\{ -\left(\omega_{RG}\right)^2_{xz} \quad K_{R_{xz}} \quad \sigma^{(4)}_{xz}(RG) \right\} s.$$

$$A_{35,5} = \left\{ -\left(\omega_{RG}\right)^2_{xz} \quad K_{R_{xz}} \quad \sigma^{(5)}_{xz}(RG) \right\} s.$$

$$A_{35,6} \cdots\cdots A_{35,19} = 0.$$

$$A_{35,20} = \left\{ -\left(\omega_{RG}\right)^2_{xz} \quad K_{R_{xz}} \right\} s.$$

$$A_{35,21} \cdots\cdots A_{35,34} = 0,$$

$$A_{35,35} = \left\{ 1 \right\} s^2 + \left\{ 2\left(\xi_{RG}\ \omega_{RG}\right)_{xz} \right\} s + \left\{ \left(\omega_{RG}\right)^2_{xz} \right\}.$$

$$A_{35,36} \cdots\cdots A_{35,40} = 0.$$

$$A_{36,1} \cdots\cdots A_{36,5} = 0.$$

$$A_{36,6} = \left\{ -\left(\omega_{RG}\right)^2_{xy} \quad K_{R_{xy}} \quad \sigma^{(1)}_{xy}(RG) \right\} s.$$

$$A_{36,7} = \left\{ -\left(\omega_{RG}\right)^2_{xy} \quad K_{R_{xy}} \quad \sigma^{(2)}_{xy}(RG) \right\} s.$$

$$A_{36,8} = \left\{ -\left(\omega_{RG}\right)^2_{xy} \quad K_{R_{xy}} \quad \sigma^{(3)}_{xy}(RG) \right\} s.$$

$$A_{36,9} = \left\{ -\left(\omega_{RG}\right)^2_{xy} \quad K_{R_{xy}} \quad \sigma^{(4)}_{xy}(RG) \right\} s$$

$$A_{36,10} = \left\{ -\left(\omega_{RG}\right)^2_{xy} \quad K_{R_{xy}} \quad \sigma^{(5)}_{xy}(RG) \right\} s.$$

$$A_{36,11} \cdots\cdots A_{36,20} = 0.$$

$$A_{36,21} = \left\{ -\left(\omega_{RG}\right)^2_{xy} \quad K_{R_{xy}} \right\} s.$$

$$A_{36,22} \cdots\cdots A_{36,35} = 0.$$

$$A_{36,36} = \left\{ 1 \right\} s^2 + \left\{ 2\left(\xi_{RG}\ \omega_{RG}\right)_{xy} \right\} s + \left\{ \left(\omega_{RG}\right)^2_{xy} \right\}.$$

$$A_{36,37} \cdots\cdots A_{36,40} = 0.$$

$$A_{37,1} \cdots\cdots A_{37,10} = 0$$

$$A_{37,11} = \left\{ -\left(\omega_{RG}\right)^2_{zy} \quad K_{R_{zy}} \quad \sigma^{(1)}_{zy}(RG) \right\} s$$

$$A_{37,12} = \left\{ -\left(\omega_{RG}\right)^2_{zy} \quad K_{R_{zy}} \quad \sigma^{(2)}_{zy} (RG) \right\} s.$$

$$A_{37,13} = \left\{ -\left(\omega_{RG}\right)^2_{zy} \quad K_{R_{zy}} \quad \sigma^{(3)}_{zy} (RG) \right\} s.$$

$$A_{37,14} = \left\{ -\left(\omega_{RG}\right)^2_{zy} \quad K_{R_{zy}} \quad \sigma^{(4)}_{zy} (RG) \right\} s.$$

$$A_{37,15} = \left\{ -\left(\omega_{RG}\right)^2_{zy} \quad K_{R_{zy}} \quad \sigma^{(5)}_{zy} (RG) \right\} s.$$

$$A_{37,16} \quad \cdots\cdots \quad A_{37,18} = 0.$$

$$A_{37,19} = \left\{ -\left(\omega_{RG}\right)^2_{zy} \quad K_{R_{zy}} \right\} s.$$

$$A_{37,20} \quad \cdots\cdots \quad A_{37,36} = 0.$$

$$A_{37,37} = \left\{ 1 \right\} s^2 + \left\{ 2 \left(\xi_{RG} \, \omega_{RG}\right)_{zy} \right\} s + \left\{ \left(\omega_{RG}\right)^2_{zy} \right\}.$$

$$A_{37,38} \quad \cdots\cdots \quad A_{37,40} = 0.$$

$$A_{38,1} = \left\{ -\sigma^{(1)}_{xz} (PG) \right\}.$$

$$A_{38,2} = \left\{ -\sigma^{(2)}_{xz} (PG) \right\}.$$

$$A_{38,3} = \left\{ -\sigma^{(3)}_{xz} (PG) \right\}.$$

$$A_{38,4} = \left\{ -\sigma^{(4)}_{xz} (PG) \right\}.$$

$$A_{38,5} = \left\{ -\sigma^{(5)}_{xz} (PG) \right\}.$$

$$A_{38,6} \quad \cdots\cdots \quad A_{38,19} = 0.$$

$$A_{38,20} = \left\{ -1 \right\}.$$

$$A_{38,21} \quad \cdots\cdots \quad A_{38,37} = 0$$

$$A_{38,38} = \left\{ \tau_{P_{xz}} \right\} s + \left\{ 1 \right\}.$$

$$A_{38,39} \quad \cdots\cdots \quad A_{38,40} = 0.$$

$$A_{39,1} \quad \cdots\cdots \quad A_{39,5} = 0.$$

$$A_{39,6} = \left\{ -\sigma^{(1)}_{xy} (PG) \right\}.$$

$$A_{39,7} = \left\{ -\sigma^{(2)}_{xy} (PG) \right\}.$$

$A_{39,8} = \left\{ -\sigma_{xy}^{(3)} (PG) \right\}.$

$A_{39,9} = \left\{ -\sigma_{xy}^{(4)} (PG) \right\}.$

$A_{39,10} = \left\{ -\sigma_{xy}^{(5)} (PG) \right\}.$

$A_{39,11} \cdots A_{39,20} = 0.$

$A_{39,21} = \{-1\}.$

$A_{39,22} \cdots\cdots A_{39,38} = 0.$

$A_{39,39} = \left\{ \tau_{P_{xy}} \right\} s + \{1\}.$

$A_{39,40} = 0.$

$A_{40,1} \cdots A_{40,10} = 0.$

$A_{40,11} = \left\{ -\sigma_{zy}^{(1)} (PG) \right\}.$

$A_{40,12} = \left\{ -\sigma_{zy}^{(2)} (PG) \right\}.$

$A_{40,13} = \left\{ -\sigma_{zy}^{(3)} (PG) \right\}.$

$A_{40,14} = \left\{ -\sigma_{zy}^{(4)} (PG) \right\}$

$A_{40,15} = \left\{ -\sigma_{zy}^{(5)} (PG) \right\}$

$A_{40,16} \cdots A_{40,18} = 0.$

$A_{40,19} = \{-1\}.$

$A_{40,20} \cdots\cdots A_{40,39} = 0.$

$A_{40,40} = \left\{ \tau_{P_{zy}} \right\} s + \{1\}.$

$B_{1,1} = -A_{1,17}$

$B_{2,2} = -A_{2,17}$

$\vdots \qquad \vdots$

$B_{21,21} = -A_{21,17}$

$B_{22,22} = B_{23,23} = B_{24,24} = B_{25,25} = 0.$

$B_{26,26} = -A_{26,17}$

$\vdots \qquad \vdots$

$B_{29,29} = -A_{29,17}$

$B_{30,30} = 0$

$B_{31,31} = -A_{31,17}$

$B_{32,32} = B_{33,33} = B_{34,34} = B_{35,35} = B_{36,36} = B_{37,37} = B_{38,38} = 0$

$B_{39,39} = B_{40,40} = 0$

$C_{1,1} = -A_{1,18}$

$C_{2,2} = -A_{2,18}$

$\vdots \qquad \vdots$

$C_{25,25} = -A_{25,18}$

$C_{26,26} = C_{27,27} = C_{28,28} = C_{29,29} = 0.$

$C_{30,30} = -A_{30,18}$

$C_{31,31} = C_{32,32} = C_{33,33} = C_{34,34} = C_{35,35} = C_{36,36} = 0$

$C_{37,37} = C_{38,38} = C_{39,39} = C_{40,40} = 0$

$D_{1,1} \quad D_{2,2} \quad D_{3,3} \quad , \cdots\cdots , D_{31,31} = 0$

$D_{32,32} = 1$

$D_{33,33}, D_{34,34}, \cdots\cdots, D_{40,40} = 0.$

$E_{1,1}, E_{2,2}, \cdots\cdots, E_{32,32} = 0.$

$E_{33,33} = 1$

$E_{34,34}, E_{35,35}, \cdots\cdots, E_{40,40} = 0.$

$F_{1,1}, F_{2,2}, \cdots\cdots, F_{33,33} = 0.$

$F_{34,34} = 1.$

$F_{35,35}, F_{36,36}, \cdots\cdots, F_{40,40} = 0$